TECHNOLOGY IN CONTEXT

T0138872

This text is a comprehensive and accessible analysis of technology assessment. The author argues against a narrow view of technology as simply a means of making a profit and instead advocates a long-term view which considers the wider interests of the firm and society.

Technology in Context defines and describes technology assessment in detail and shows the role it can and should play in the strategic management of firms. In particular, the author demonstrates what information is needed to manage technology effectively and with the greatest possible benefit to the firm and its employees. Technology assessment supplies an early warning of the potential hazards of technologies, and provides the necessary information to allow managers to make informed choices about the impact of technology on the firm and society. The text also explores the main problems associated with the application of technology including the danger of environmental degradation and the problems of employment and skills.

Technology in Context is a complete introduction to the theory and methods of technology assessment as a tool of strategic management. It will be a useful book for all those interested in the management and social role of technology.

Ernest Braun is Emeritus Professor and former Head of the Technology Policy Unit, Aston University, UK. He has also held the posts of Head of the Technology Assessment Unit, Austrian Academy of Sciences, Vienna, and Visiting Professor, Centre for Technology Strategy, Open University. His books include *Futile Progress* (1995) and *Wayward Technology* (1986).

THE MANAGEMENT OF TECHNOLOGY AND INNOVATION

*Edited by David Preece, University of Portsmouth, UK,
and John Bessant, University of Brighton/SPRU, UK*

The books in this series offer grounding in central elements of the management of technology and innovation. Each title will explain, develop and critically explore issues and concepts in a particular aspect of the management of technology/innovation, combining a review of the current state of knowledge with the presentation and discussion of primary material not previously published.

Each title is designed to be user-friendly with an international orientation and key introductions and summaries.

New titles in this series include:

TELEWORKING: INTERNATIONAL PERSPECTIVES

From Telecommuting to the Virtual Organisation
*Edited by Paul J. Jackson, Brunel University, UK,
and Jos M. van der Wielen, Tilburg University, The Netherlands*

TECHNOLOGY IN CONTEXT

Technology assessment for managers

Ernest Braun

London and New York

First published 1998
by Routledge
2 Park Square, Milton Park, Abingdon, Oxon, OX14 4RN

Transferred to Digital Printing 2004

Simultaneously published in the USA and Canada
by Routledge
29 West 35th Street, New York, NY 10001

Typeset in Garamond by Routledge

British Library Cataloguing in Publication Data
A catalogue record for this book is available from the British Library

Library of Congress Cataloging in Publication Data
Braunm Ernest, 1925–
Technology in context : technology assessment for managers /
Ernest Braun.
Includes bibliographical references and index.
1. Technology assessment. 2. Technology–Management. I. Title.
T174.5.B75   1988
657.5'14–dc21   97-37967
CIP

ISBN 0–415–18342–1 (hbk)
ISBN 0–415–18343–X (pbk)

CONTENTS

FIGURES

INTRODUCTION

Technology can be used for good or ill. How it is used depends largely on how it is managed, and to manage it well we need to look upon it with as broad and far-reaching a perspective as possible. To take a hard, broad and long-range look at technology is the essence of both good management and technology assessment.

The present text aims to serve students of technology management. After setting out some of the fundamental concepts required for the effective management of technology, the text aims to treat in detail one aspect of the problems facing managers of technology: the gathering of information to support strategic decisions on technology. The book aims to show what information is needed to manage technology effectively and with the greatest possible benefit to the firm and those who work in it or interact with it. I believe that the information should be wide-ranging and far-sighted and go well beyond the simple reading of a few newspaper or journal articles and sales brochures. Indeed it is my belief that the more knowledge is obtained before a decision is made on the development or deployment of a technology, the greater the chances that fewer mistakes will be made and fewer unpleasant surprises will be encountered. In particular, I believe that technology should be viewed in its widest context, that is in all its impacts upon society, the natural environment, and the organization it is deployed in. This kind of information gathering is known as *technology assessment*, and that is what this book is all about. The purpose of technology assessment is to look beyond the immediately obvious and analyse the ramifications of a given technology in as wide-ranging and far-sighted a manner as possible. I believe that only by gathering the best possible information can we make the best possible decisions. Technology assessment is a systematic attempt to remove all blinkers and to cure myopia.

1

My ultimate goal is to make a contribution, however modest, toward a wiser and better use of technology. A use that obtains the undoubted benefits from technology without paying severe penalties in terms of environmental destruction, health hazards, de-skilling, unemployment and an inhuman ambience.

I have endeavoured to make the book both informative and enjoyable – my readers will be the judges of my success. I do not believe that these aims would be achieved by producing a text of erudite theory. I believe in a practical down-to-earth approach, with enough theory to put matters in a wider context and to provide some intellectual stimulus.

My sincere thanks are offered to the many colleagues and students who have contributed so much to my knowledge and to my thinking. I wish to thank the Centre for Technology Strategy of the Open University, and particularly David Wield and Mary McVay, for their active support of this work. My special thanks are owing to John Bessant, who gently persuaded me to undertake this project. Personal thanks are further owing to Stuart Macdonald, Walter Peissl and Helge Torgersen for their help and support. Finally, I wish to thank my wife, Dorothea, for her encouragement and for putting up with much inconvenience to enable me to complete this task.

Outline of the plan of the book

The first chapter sets out to define and describe some of the main concepts used throughout the book. It starts by justifying briefly why we need specialist technology managers and goes on to define technology at some length. This is a controversial topic, and I try to establish some common ground, at least for the purpose of this book.

I then define technological innovation and describe some aspects of the theory of innovation, because I think that this is essential background to the understanding of technology management and of technology assessment.

The second chapter starts with a definition of technology assessment (TA). It then describes the beginnings of technology assessment as a response to problems of technology policy formation in the United States Congress. Though TA started in the public domain, many of its principles and features can usefully be transferred to use by commercial firms. The most general methodology used in all technology assessments is described in some detail and the fundamental difficulties and controversies faced by technology assessment are discussed. The chapter concludes with a very general example of a technology assessment, the assessment of modern telecommunications, to illustrate the use of the methodology.

Because one of the principal features of technology assessment is the attempt to look as far as possible into the future, the main beneficiary from technology assessment in commercial firms is the long-term, or strategic, management of technology. Hence the third chapter provides an outline of the fundamentals of strategic management in general, and of strategic management of technology in particular. The role of technology in commercial firms is discussed in some detail, including its role in the strategic positioning of firms. Finally, the formation of technology policy in the public and commercial domains are compared and the role of technology assessment as an information input to policy formation is described.

The need for strategic management of technology, and the need for comprehensive far-sighted information, stem from the fact that technology has become one of the central determinants of economic success and of the way we live and work (or do not work). Social organization and technology are inextricably entwined. Technology is also seen to have caused many worrying environmental problems and technology assessment is in part an attempt to prevent future negative impacts of technology upon the natural environment. Hence the fourth chapter is devoted to reviewing the contemporary problems, fears, and hopes linked to technological developments. Among the hopes I count the opportunities for creating socially useful new technologies.

The fifth chapter discusses some of the methods useful in technology assessment. Among these, forecasting is discussed in some detail as it invariably is an aspect of technology assessment. Clearly we cannot know the future, yet everything we do shapes the future and hence forecasting plays an important role in formulating strategies for technology. As technology is an important determinant of economic performance, economic analysis of the impact of any given technology inevitably forms part of technology assessment. Similarly, methods for assessing environmental impacts are included. The methods are illustrated by some examples.

The sixth chapter deals with further examples of technology assessments. Environmental impact assessment and planning applications are important special cases of the general approach. These are important tasks of strategic analysts that are very closely related

3

to technology assessment proper. Fundamental considerations underlying decisions on the location of new plant are briefly mentioned in this context.

Two further examples of technology assessments are given: an assessment of agriculturally produced renewable raw materials, carried out in the public domain in Germany; and an imaginary assessment of solar heating panels, supposedly carried out by a small firm in a Southern European country.

The chapter concludes with remarks about TA in the field of biotechnology and with a few general remarks on technology assessment.

1

TECHNOLOGY AND TECHNOLOGICAL INNOVATION

Technology management

Technology has become a vital part of the life not only of commercial and industrial firms, but of the lives of individuals and societies. Whether we speak of materialist society or of information society, it is technology that lies at the heart of producing material wealth and we use technology to handle, store and transmit information. We use technology for better or for worse and it is the task of technology management to tilt the balance toward the better.

This text is aimed at managers of technology. Technology makes specialist demands upon managers that cannot readily be met by generalist managers or by engineers without training in the management of technology. There are complexities and peculiarities about the management of technology that demand specialist knowledge and specialist approaches. Whereas the application, creation, design, maintenance and improvement of technology itself are, of course, the domain of the engineer and scientist, managing technology in the context, and for the benefit, of a firm is the domain of the technology manager. To paraphrase the famous dictum by Clemenceau about war being too important to be left to the generals: technology is too important to the life of firms to be left to technologists.

The task of management of a firm consists of coordinating and controlling its activities so as to serve its best interests. Leaving aside the question of what a firm's best interests are and who determines them, the task of managing consists of many interrelated functions. Clearly, there is general management, usually at the hierarchical pinnacle of the firm. At this level, the firm's strategy is decided and its different functions are coordinated and controlled. Each separate function of the firm also needs to be managed. Thus we usually speak of financial management, personnel management, sales management, and so forth. But surely if technology is such a vital aspect of a firm's activities, technology management must also be an essential task. Indeed technology is managed, generally under titles such as production management, R&D management, communications management. All these functions are specialist aspects of technology management. In information-intensive service organizations, such as banks, technology management usually falls to computer and communications managers. The term technology manager is an umbrella concept covering all the varied aspects of the management of technology.

It is common for technology management functions to be fulfilled by technologists who learn on the job. This can work well, and indeed there is no substitute for learning on the job, but learning on the job is usually easier and more effective on the basis of some theoretical foundation.

There are three types of enterprise likely to need specialist technology managers:

1   Manufacturing firms. These need, first and foremost, to manage their manufacturing technology (process technology). To some authors, this is the very essence of technology management. If the firm produces a complex product, then the task of product development also becomes a technology management task. It goes without saying that product and process technology are inextricably interwoven and that their management must be closely coordinated. In many cases the production of a new product demands new production technology and, more important, the design of a new product must bear ease of production in mind. Ease of manufacture, so-called 'manufacturability' is a key to major cost savings.

2   Service providers. Many service providers, such as insurance companies, transport undertakings, hospitals, or even retailers, now use highly sophisticated technology, ranging from computers to telecommunications and to automated warehouses. For some service providers, such as airlines or railways, specialized complex technological systems are at the very core of their business. Purchasing and maintaining aircraft and computer and communications systems are among the most important activities of an airline. Hospitals use a wide range of expensive diagnostic and therapeutic technologies. The demands of management of technology in such organizations

6

are complex and distinct from the requirements of other specialist or general management.

3   Utilities, extraction industries, construction, etc. Clearly, all these types of firm use complex technologies that are their life-blood, their raison d'être. It should not surprise us that the management of these technologies is important and requires the attention of specialist managers.

At a higher level of generalization, almost any commercial firm has two sides to it: producing a product or service, and selling it. Technology may or may not be used on the sales side, but it almost invariably lies at the heart of production. Each side, and its component parts, needs to be managed, and the whole needs to be purposefully coordinated.

The present text is not a general text on technology management; it deals merely with one aspect of the task: technology assessment. We may sum up this task by the proverb 'look before you leap'. In other words, think long and hard about the consequences of choosing a certain technology before deciding on it. Technology assessment is neither more nor less than the task of gathering sufficient information about a technology, and its likely future consequences for all those who interact with it, before embarking upon developing or deploying this technology. This sounds like simple common sense and so it is, but the application of simple common sense requires a great deal of knowledge and specialized techniques. The knowledge that is obtained as a result of much thought and investigation by many well-trained people is deeper, and more dependable, than easy conclusions that look like common sense knowledge at first sight. True knowledge is built up gradually and painstakingly – it does not appear in a flash of inspiration, even if inspiration has an important role to play in the creation of knowledge.

What we aim for in this text is to help the student to obtain a clear grasp of what sort of information is required for wise decisions about technology, and indeed what distinguishes wise from unwise in the technological context. We shall not lay down rules of wisdom, but attempt to provide some guidance for the student's own thoughts. When all is said and done, many aspects of wisdom are a matter of values and judgement and cannot be uniquely determined. What action is wise depends on what values you cherish and on what you wish to achieve.

Before we embark upon the details of technology assessment, we need to define and clarify some basic concepts. We start with technology.

The definition of technology

We all think we know what technology is, but when we think about it more closely, the term becomes elusive because it is so difficult to distinguish technology from other types of human activity. Definitions are important not only because they define and describe a subject, but also because they exclude other subjects. Thus definitions set boundaries for a topic of discussion, to the exclusion of all other topics. Set the boundaries too narrowly and the discussion becomes sterile, scholastic and pointless; set them too widely and the discussion becomes diffuse, unfocused and nebulous.

Definitions of such potentially all-embracing topics as technology cannot be unique and universally accepted. They must depend not only on the personal taste and attitude of the author of the definition but, more important, on the purpose the definition is to serve. In our case, in a textbook for technology managers, the definition of technology must be aimed at serving the needs of this particular audience. The role of the technology manager is a distinct one, different from the roles of other types of manager, because technology is a distinct class of objects, knowledge and activities. By defining technology, we define the domain of the technology manager. Each author on technological issues is forced to produce an own definition of technology, or select one from a multitude that have been proposed. We prefer to give our own definition, though without raising the claim that it is the only, or even the best possible, definition.

If we define technology too narrowly, we constrain ourselves to speaking of machines and tools. Defining technology too widely, on the other hand, perhaps as the organization and method for the production of material wealth, does not allow us to distinguish between technology and other purposeful activities, such as commerce, marketing, law, or accountancy. One such all-embracing definition is: 'Technology means the systematic application of scientific or other organized knowledge to practical tasks' (Galbraith 1974, 31). This would make the psychotherapist a technologist and that, to my mind, is casting the net too wide.

We define technology as *the ways and means by which humans produce purposeful material artefacts and effects*. Alternatively, we may define technology as *the material artefacts used to achieve some practical*

8

human purpose and the knowledge needed to produce and operate such arte-facts. Essentially, in this definition technology always consists of material artefacts (hardware, means), and of the software (knowledge, ways) necessarily and immediately associated with it.

An acceptable alternative definition of technology is: 'A process which, through an explicit or implicit phase of research and development (the application of scientific knowledge), allows for commercial production of goods or services' (Dussauge, Hart and Ramanantsoa 1992, 13). These authors use the term 'technology' as a high-level concept, whereas they prefer the term 'technique' for more mundane, less abstract, usage. This comes naturally to French or German speakers, as the French word *technique* (*Technik* in German) has a broader meaning in French (or German) than in English.

In broad terms, we distinguish between the product of a firm and the technologies employed to produce it. Thus we should distinguish between technology itself and products of technology, though some of the latter may themselves be technological devices. We shall, however, not insist on this distinction and follow the common practice of distinguishing only between production (or process) technology and product technology. The product of one firm may well be the production technology of another firm; for example the output of a machine tool manufacturer is used by other manufacturers to manufacture a range of various products. Though the basic, and commonly used, classification of technology knows only two categories: process (or production) technology on the one hand, and product technology on the other, it is sometimes useful to make more detailed distinctions between different types of technology. This further clarifies the nature of technology and gives a first inkling of the different tasks involved in managing different types of technology. The following might serve as a useful, though incomplete, classification:

•Production technology, i.e. purposeful systems of tools and machinery used to produce a variety of products. A system of production is not a random collection of suitable machines, but a system designed for a purpose, incorporating machines linked, controlled and managed in sophisticated and complex ways. We regard the organization of the system and the various measurement and control functions incorporated in it as part of the system. The system generally consists of many sub-systems, including quality control, stock control, materials handling, and so forth. Examples of such systems are the array of machinery, tools and instruments used to produce automobiles; or the machinery and control systems used to produce chemical fertilizers; or the somewhat simpler system of machinery used to produce shoes.

Sometimes a distinction is made between process and production technology, though more often these terms are used as synonyms and we shall not insist on the distinction. Strictly speaking, a process is something like a chemical process, a fermentation process, or welding, or turning. Production technology uses a variety of processes in a system to produce a product.

•The products of technology are many and varied. We may usefully distinguish between several categories.

1   Technological implements, i.e. tangible artefacts used to achieve some desired practical purpose or effect. We use this term to include tools and implements such as hammers, ploughs, or pots and pans. It may seem a little far-fetched to call furniture, garments and footwear technological implements – but though their use is more passive, the definition still fits them.

2   Measuring instruments and control devices, such as micrometers, thermometers, or strain-gauges. Measuring devices and sensors often form part of a manufacturing system or of more complex devices, such as vehicles.

3   Devices that use energy to achieve some physical effect, such as air conditioners, refrigerators, heaters, furnaces, lighting.

4   Vehicles, such as cars, railway engines and carriages, aeroplanes. These could be subsumed under point 3, but are too important not to be given a classification of their own.

5   Devices used to achieve effects that are not primarily physical. This category includes the whole range of devices used in information and communication technology and in entertainment, such as computers, telephones, video-recorders.

6   Engines, motors and machines. Engines and motors convert some form of energy into mechanical motion; machines use mechanical motion to achieve some desired effect, such as turning, washing clothes, pumping water. Robots and conveyor belts also fall into this category.

7   Building and construction, including houses, roads, bridges, dams, tunnels and so forth.

8   Processing technologies, such as the production of engineering and other materials, chemicals, pharmaceuticals, food, and so forth.

•Technological systems, e.g. complex sets of devices and material artefacts, serving some practical material purpose.[1] An example is the railway system, consisting of track, bridges, a signalling subsystem, stations, locomotives, rolling stock, and an elaborate set of rules and procedures to enable the system to function as an effective carrier of goods and passengers. Our definition of technology includes those rules and procedures that are directly related to the operation of the technical system, e.g. the rules controlling activities of engine-drivers, or procedures related to the maintenance of rolling stock. We exclude matters such as financial management, ticket sales, or catering.

A road is part of a technological system; the system of road transport. The boundaries of this system can be chosen somewhat arbitrarily, but we may sub-divide it into the technological system, i.e. roads, bridges, traffic lights, petrol stations, the system of fuel supply, cars and lorries; and a secondary system, consisting of laws, law enforcement, emergency services, etc. Strictly speaking, ambulances and medical services are not necessary for the functioning of the road transportation system, but it is hard to envisage a society that does not try to ameliorate some of the horrors that are the by-products of their choice of technological system.

To some people, technological knowledge itself is technology. This is debatable, but certainly the knowledge necessary to design, construct and operate technological artefacts cannot be properly divorced from the artefacts themselves. Thus the knowledge necessary to design and produce automobiles, and a production system to manufacture them, is technological knowledge and is a pre-condition for the manufacture of automobiles. Generally speaking, technological knowledge is the knowledge dispensed in engineering courses. It is a moot point whether the knowledge dispensed in a course on the management of technology is technological knowledge. On balance, we tend to think that some of it is, but much of it is not because it attempts to consider non-technological factors that are important to the management of technology.

The word technique, although closely related to the word technology, is used in a narrower sense to describe the way we perform certain tasks, e.g. the technique of using a tool, or a

---

1   A telephone system, consisting essentially of hardware, is a technological system, even if it is debatable to what extent its effects are material.

measurement technique. Technique is akin to method, manner of execution, or skill.

Purely organizational arrangements, where the use of physical artefacts is trivial or not of the essence, are not regarded as technology. Thus a system of sales outlets and representatives, however vital to the operations of a manufacturing firm, is not considered technology, though the sales force may use technology for the performance of certain tasks, such as communications, control, or display. The general organization of a firm should not be regarded as technology, any more than a system of governance or the organization of a church should be considered as technology.

With the advent of computers and their ever widening use, it is tempting to widen the definition of technology to include all computer software. Indeed the computer, including its systems and applications software, is a technological device par excellence, and computer software may be regarded as technology in so far as it is inseparable from a computer. Strictly speaking, we must consider a book as a product of technology, albeit a passive one. We must beware, however, of identifying the computer with the tasks it can perform. Even if a balance sheet is produced by computer, this does not make accountancy into a technology. We must maintain a boundary between technology and the purposes for which it is used, otherwise everything becomes technology. After all, music uses instruments which are products of technology, but musicians are not technologists; linguistics uses computers, but linguistics is not a technology; and even the dreaded inspector of taxes is not a technologist, despite the widespread use of computers in tax offices. We must not fall into the trap of confusing the tool with the task performed by it.

Our definition of technology encompasses the hardware, the tangible artefacts, used to perform some practical task, as well as the software immediately associated with the hardware. Although technology and organization interact strongly, we regard a distinction between them as useful. The efficacy and efficiency of a production system is strongly influenced by its organizational features, but organization without machinery cannot produce any artefacts and should not be regarded as technology. The finance director and the chief engineer of a firm do perform distinctly different roles.

In many ways medicine might be regarded as a technology. Our definition, however, sticks to the convention of regarding medicine as a non-technological empirical mix of art and science, though in modern times heavily dependent upon the use of sophisticated, specialized and expensive technological devices.

Technological innovation

Homo faber is a restless creature; never satisfied, never still. One of the most characteristic features of technology is the fact that it is undergoing constant change. Technology never stands still, never reaches a steady-state equilibrium. Two major forces push technology toward change. One is the internal logic of science and technology. Technology is never so perfect that it could not be improved, and new knowledge brings with it new technological possibilities. The second force is economic. Technology is a crucial weapon in the relentless quest for economic growth. New technologies and new products on the market expand economic activity; improved efficiency of production increases wealth, provided society can absorb it without increases in unemployment. Indeed, current theories of economic growth emphasize the role of increasing productivity, caused by capital investment into new and more productive machinery, the role of R&D in producing new technologies, and the role of better education and training in technology and management. Technology and its efficient use have come to take pride of place among the ingredients that determine competitive advantage and economic success of nations and of enterprises (OECD 1992, 167–185).

Technology is wont to change in two directions: toward higher performance, and toward the satisfaction of an ever increasing range of potential human wants (Braun 1995, 20–29).

Higher performance of technology means higher speed, larger capacity, improved cost-effectiveness, higher efficiency, increased reliability, reduced effort required from human labour, greater comfort in use, more automated features. Production machinery becomes faster, more automatic and capable of carrying out more tasks; cars become faster, more comfortable, need less servicing and, ideally, become safer; ships become bigger and need smaller crews; computers can carry out more and more operations per unit time and can store more information. Telecommunication equipment can switch faster and carry more information per line and unit time. Central heating requires no human effort to keep a house at a much higher level of comfort than the labour-intensive, inefficient, dirty (but nice) open fire.

In trying to satisfy ever more potential human wants, technology takes over more and more tasks formerly carried out by humans, and extends the range of activities and possibilities available to humans. The washing machine or the pocket calculator take over chores previously performed by humans; the video-camera or the aeroplane enable humans to indulge in activities that are inaccessible to them without the aid of technology. Similarly, the telephone, television, the compact disc player, the computer, the freezer, and many more implements extend the range of things we can do. Technology serves as both crutch and magic wand.

Technological innovation needs to be distinguished from invention. An invention is a novel technological idea that need never reach production or the market. A technological innovation is a new, or substantially improved, technology, or product of technology, that is offered for commercial transactions on the market.

The process of technological innovation is held to be of great importance and, hence, is the subject of much theoretical analysis. Theories of technological innovation concentrate on three aspects: the process of technological innovation and its various stages; the classification of innovations according to technological and economic criteria; and the management of innovation.

We shall deal with the taxonomy of innovations first. They are commonly classified in three dimensions: the type of technology involved, such as product innovation or process innovation; the degree of technical change, distinguishing mainly between incremental and radical innovation; and the extent of the socio-economic influence of the innovation. Freeman and Perez, whose main interest lies in macro-economics, combine criteria of degree of novelty with criteria of socio-economic importance into the following, widely accepted, classification into four classes (Freeman and Perez 1988, 38–66).

First, incremental innovations. These are innovations that do not involve a great leap in technology. They occur continuously in all types of technology. They are not necessarily the result of deliberate R&D, but of re-design of existing products or design of new products in response to discovered weaknesses or to good ideas, sometimes stimulated by users. The cumulative importance of these innovations is quite substantial, both at the level of the firm and at the macro-economic level.

Second, radical innovations. These involve radically new technical or scientific ideas. They occur less frequently and are usually the result of deliberate R&D. Examples are nylon, or the transistor. These innovations, if successful, may create entirely new markets,

stimulate stagnant markets, and – in the case of production technology – may cause major improvements in productivity. They may
afford new investment opportunities and may be associated with
some organizational and skill changes. If a cluster of radical innovations are linked together, they may give rise to whole new
industries, such as the semiconductor industry (Braun and
Macdonald 1982).

Third, changes of 'technology system'. This type of change is
based on several radical and incremental innovations and has far-
reaching effects on more than one branch of industry. It gives rise to
new industrial sectors and causes managerial and organizational
change in addition to technical change.

Fourth,

> Changes in 'techno-economic paradigm' ('technological
> revolutions'): Some changes in technology systems are so
> far-reaching in their effects that they have a major influence
> on the behaviour of the entire economy. A change of this
> kind carries with it many clusters of radical and incre-
> mental innovations, and may eventually embody a number
> of new technology systems. A vital characteristic of this
> fourth type of technical change is that it has pervasive
> effects throughout the economy, i.e. it not only leads to the
> emergence of a new range of products, services, systems and
> industries in its own right; it also affects directly or indi-
> rectly almost every other branch of the economy . . . the
> changes involved go beyond engineering trajectories for
> specific product or process technologies and affect the input
> cost structure and conditions of production and distribu-
> tion throughout the system . . . once established as the
> dominant influence, [it] becomes a 'technological regime'
> for several decades.
>
> (Freeman and Perez 1988, 47)

The main thrust of the above classification is the attempt to establish technological causes for major socio-economic upheavals,
dividing the industrial epoch into distinct periods, each dominated
by one, or a few, major technologies. Sometimes these periods are
regarded as long waves, named after the Russian economist
Kondratiev. Each complete wave consists of a rise toward prosperity,
supposed to be associated with a particular cluster of technologies,
followed by a decline, when the main economic potential of the

particular techno-economic paradigm is exhausted. The four previous periods are associated with a cluster of textile innovations and steam power (1770–1830); railways and steel production (1830–1880); electricity, the internal combustion engine and the chemical industry (1880–1940); pharmaceuticals, plastics, television, and electronics (1940–1980) (Freeman, Clark and Soete 1982, 66, 127; and Bessant 1991, 5).

It is widely believed that we are currently in the early stages of a 'fifth wave' of techno-economic paradigm shifts. The present paradigm is dominated by microelectronics, computers, telecommunications, and, possibly, bio-engineering. It is also characterized by a shift from the Fordist paradigm of mass-production to a new post-Fordist paradigm. Whereas the Fordist paradigm was based on the energy-intensive mass-production of a rather stable mix of products, specialized skills, and hierarchical management structures, the new paradigm is supposed to be information-intensive, with flexible production systems, multi-skilled workers, and flat management structures. The exact shape and features of the new techno-economic paradigm have not become clearly discernible. It appears that we are in a period of flux, with great emphasis on flexibility, adaptability, organizational learning, and a general lack of stability. It is often emphasized that price alone is no longer decisive in the competitive position of many goods, as design, technical features, prompt delivery, after sales service and similar so-called non-price factors play an increasingly important role.

To the technology manager these matters are important from two points of view. First, he/she needs to know which way things, in general, are moving. There is a Zeitgeist about technology and its applications and, though it is possible to tread different paths, this needs to be done, if at all, knowingly and with conviction.

This is a time when competition is tough, and when political thinking has returned to blessing the successful and forgetting the unfortunate. There has been a great shift of power and income from workers – even skilled workers – to top managers, owners and manipulators of capital, bankers, and what might be called an 'expertocracy' (highly paid consultants and top experts). Though technological changes have probably contributed to and enabled these changes in social structure, there are other, purely political/economic forces at work as well.

Beware, though, of euphemisms. To call this society 'active' because people have to live on their wits, to call it 'flexible' because people no longer have stable full-time jobs, and to call the act of throwing people out of jobs 'rationalization' or 'down-sizing', are all euphemisms designed to obfuscate rather than clarify, to make the pill more palatable without reducing its bitterness. The good manager needs to know things as they are, not as they are presented by PR people, politicians, or the media. And do not blame it all on technology!

Second, macro-economic theories of innovation matter to the technology manager because they explain concepts such as techno-economic paradigm, technological regime, and technological trajectory. Though they are often couched in complex economic terminology, these concepts help to remove some uncertainty in that they show that technological developments tend to follow rela-tively ordered paths. Knowing these helps to ask the right questions and instigate the right searches, thus allowing a degree of foresight and a degree of confidence. The latter is particularly necessary when investors and senior managers are asked to spend large sums of money on technological innovations or investments. If a technology is developing on a reasonably predictable path, if a technological regime is likely to last, then investment in this technology appears much safer (Dosi and Orsenigo 1988).

There always are many more ideas for technological innovations than are ever fully developed and marketed. There is some form of social selection mechanism – a kind of social Darwinism – that decides which ideas will thrive and which will wither. These mecha-nisms operate within firms, where ideas for development are sifted and whittled down. They operate even in the earliest stages of research, where support is given highly selectively. And they operate in the market, where some products and services fail to gain accep-tance. Once a technology is generally adopted and well-established on an ordered path of incremental progress – a technological trajec-tory – competing ideas have little or no chance. Innovations in this field then occur along the trajectory, or in technologies that support or complement the main technology, thus contributing to the formation of a technological system. Once, for example, silicon chips and digital electronics became fully established, competition from potential rival technologies virtually ceased.

Though technology managers should be aware of macro-economic trends, their domain and sphere of activity is the micro-economic one. Hence a few remarks on the micro-economic theory of innova-tion are called for. These theories are mainly concerned with three aspects of the process of innovation: types of innovation; sources of

17

innovation; stages of innovation. And, of course, the best possible management of innovation.

For micro-economic purposes we classify innovation according to two criteria only. *On the one hand, we distinguish between product and process innovation; on the other hand, we distinguish between radical and incremental innovation.*

Process innovation includes the development of new production machinery and new production processes, but it also includes what has been termed *manufacturing innovation, i.e. the introduction of new production methods into a plant*, even if the methods were bought in and are not entirely new. By product innovation we mean mainly the development of new or radically improved products for final consumption, i.e. mostly, but not exclusively, non-capital goods. The distinction is not always a clear one, for when manufacturers of, say, textile machinery introduce a new type of machine, this is a new product to them, but a process innovation for textile manufacturers who buy the machine.

The second criterion is the degree of technical/scientific novelty of the innovation. An incremental innovation contains little novelty, whereas a radical innovation departs considerably from established practice. Normally, the radical innovation emerges from deliberate systematic R&D carried out in a laboratory, whereas incremental innovation is often the result of re-design carried out without laboratory R&D. Even these apparently obvious distinctions are not always helpful. Take the example of increasing the number of active elements in an integrated circuit, or the development of a new diesel engine. In each case one might jump to the conclusion that the innovation is incremental and easy, as no new scientific or technological principles are involved, yet this conclusion is entirely wrong. The increase of elements in an integrated circuit may require new manufacturing processes and demands a great deal of R&D; similarly, the development of a new diesel engine is an extremely laborious and costly process. The reason in each case is that we are dealing with a highly developed technology, where each further improvement is hard and expensive to obtain – this has been termed the law of diminishing returns on R&D (Braun 1995, 64–75).

The macro-economic impact of an innovation can only be ascertained with hindsight. Even radical innovations may not have a major macro-economic impact. Think, for example, of the hovercraft. On the other hand, the impact of the transistor, the integrated circuit and the microprocessor exceeded the wildest dreams of their creators.

The stages of development of an innovation may be described, in simplified and schematic form, as follows:

1   the idea or invention that emerges from a selection process;
2   development, which may involve a great deal of R&D;
3   prototype, which may need many versions and more R&D;
4   production, which may need much investment and further development;
5   marketing and diffusion.

The description of the stages of innovation seems simple and obvious, yet the stages are not necessarily sequential in time. They represent a logical, rather than a temporal sequence, and many feedback loops exist between the stages. Particularly if we are dealing with a major innovation, the timing between various aspects of the research and development work is as important as it is difficult, because outcomes of R&D are always somewhat unpredictable.

The sources of innovation are varied and the emphasis on different sources is controversial. What is not controversial is that *an innovation is the result of the confluence of a new technical idea with a market opportunity*. The controversy is about the extent to which the market actually calls for a particular innovation (so-called market pull) and the extent to which the sponsors of the innovation merely hope that their technology will find a market (technology push). Thus the source of an innovation may be the market, signalling its desires through the manufacturer's sales personnel or other contacts between producer and consumer. The source may be the firm's R&D laboratory, or suggestions by employees, or ideas that flow from the work of engineers and designers, or, now more rarely, the work of a lone inventor.

Innovation is an expensive and risky business, yet a firm that does not innovate may well go under because of the pressure of competition and because markets tend to saturate and need re-kindling with new products. For a variety of reasons – the operation of the law of diminishing returns, the great costs and risks of innovation, political pressures – it has become very common for firms, even rival firms, to cooperate in innovative projects. This makes life harder for the managers, but the increased cost of management is outweighed by the savings in costs and reduction of risk owing to sharing.

There are several aspects of technological innovation that need careful management, quite apart from the management of R&D, which is a separate issue that is only touched upon in this text. The preliminary step for any innovation is for the idea to gain acceptance. There always are more ideas for innovations than are selected for development into actual innovations. Most firms have an established procedure for this sifting and selection process. Once an idea is accepted, it is likely that a manager (or a management team) will be given responsibility for progressing the innovation through its stages. Assuming that it is at the development stage in the R&D laboratory, the first task is to nurture the innovation with a combination of foresight, hope and money through these difficult, early, but probably protracted, steps. At the same time, the manager must prepare the ground for the next stages and, if necessary, see to it that some of the work of the next stages is undertaken. The manager of the innovation must act as its champion, for without a champion within the firm the delicate plant of an innovation in its early stages is unlikely to survive. Mortality of ideas is high, but mortality of innovations in their early stages — a kind of infant mortality — is also substantial.

If the product innovation is not too radical, i.e. it does not depart too much from established practice and experience, then the idea for a new product may come from some other source, such as an internal suggestion, market research, or routine up-dating. Indeed many new products are simply the result of a re-design without much technical change and without anything that could rightfully be called R&D. Some new products are based on changes in fashion and are new more on the outside than in their workings. There is another class of product innovation based on relatively simple — yet often ingenious — invention and a little trial and error. Though these types of product innovations do not have the glamour of emerging from R&D laboratories, their cumulative effect in the economy is very important and they may play an essential role in the life and prosperity of a firm. The task of management in these cases is careful screening and selection of ideas and the coordination of support for the progress of the innovation to the next stages of the process.

The next stage of the innovation process consists of building prototypes and developing these to a finished design. The design must be functional, aesthetically pleasing, have the required technological capabilities and properties, and be economic to manufacture in the desired numbers and within the desired price limit. It may well be necessary to pose further questions to the R&D laboratory before a satisfactory prototype is obtained.

The production facilities for the new product need to be prepared while it is still at the prototype stage. New production processes may have to be developed or bought. In many instances, product and process technologies are closely intertwined. A prime example is integrated circuits (silicon chips), where design of the product and design of the process are ingredients of equal importance in any new device.

The early production runs are bound to be fraught with difficulties and it is only with accumulated production experience that productivity rises – this is the well-known learning curve. Productivity rises as a function of the numbers of the new product produced, but this is not an automatic process, it may require re-design of both product and process. Though some learning consists of the acquisition of manual dexterity and tacit knowledge, much consists of drawing deliberate conclusions from mistakes – a process that needs careful management.

The management of an innovation in process technology is very similar to the management of product innovation. If a process innovation – we use this word for brevity to include innovation in production machinery – is radical, then again it will probably emerge from a long period of gestation in the R&D laboratory. It will also require further development in prototype form – probably in a scaled-down version in the first instance. Much of this prototype work can now be done by simulation on high-powered computers, and this saves a great deal of time and effort. If the process innovation is for sale, rather than for internal consumption, then even the late stages of the process of innovation are very similar to those of product innovation. After all, one manufacturer's product is another manufacturer's process. The difference between product and process innovation is that, with the latter, final development is often carried out in conjunction with a pioneer customer, and that generally the customer is more discerning and more knowledgeable than the general public.

All types of technological innovation are of vital significance for the strategic positioning of the firm and, even more important, for its competitive strength. Product innovation is essential, for without products of competitive quality and design, sold at a competitive price, no firm can survive in the long run. But why, it may be asked, should we need to create new products, rather than continue to sell the old ones? In some cases, this is a valid question.

21

Cement remains cement, beer remains beer, flour remains flour. Yet even in these cases patterns of consumption often change. In Britain, for example, there is more demand for Pilsen type beer than there used to be and stone-ground wholemeal flour has become newly fashionable.

Apart from the constant need to maintain some competitive advantage in the product range offered by a firm, the main reason why manufacturers attempt to bring new products onto the market is fear of stagnation. When a product is entirely new, say black and white television in its beginnings, manufacturers can hope to sell as many sets as there are households. Once everybody has a black and white television set, there remains only a small replacement market and second set market. But the advent of colour television turned a saturated market into an entirely untouched one. The whole game started from scratch as gradually every household acquired a colour set. And now, when that market is saturated, hopes are pinned on digital television, high definition television, television combined with computers, interactive television, and other goodies that might revive a stagnant market.

Television is an example of a series of radical innovations. But other products undergo continual change in order to maintain demand, and because individual manufacturers must keep abreast, or ahead, of their rivals. Some of the innovations are technological, but many changes are purely stylistic. Even very complex high technology products, and motor cars are a prime example, are subject to fashion in looks, equipment and performance.

A firm may decide to be a product leader or a follower; it may decide to embark single-handed on a radical product innovation or it may enter an alliance with a competitor; it may specialize in constant up-dating; it may position itself in different market segments. All these are possible strategic positions, to be discussed in the next chapter, and their implementation lies largely in the hands of technology managers.

The benefit of successful innovation can be great, but the risks can also be substantial. A firm that is first with a truly successful innovation that is difficult to imitate can reap monopoly profits for some considerable time. On the other hand, if the innovation fails, either for technical or for commercial reasons, the losses can be very large indeed. If a firm abstains from innovation altogether, it may not run any risks, but it still incurs the penalties of losing market share or facing a saturated market.

It has been argued that too fast a pace of innovation is as harmful as too slow a pace, as the cost of innovation has to be given sufficient time to pay for itself and to reap profits. There is a careful balance to be struck. To innovate too much or too fast can be as ruinous for the firm as to innovate too little and too late. For society at large, a fast pace of innovation, without proper regard for the needs of environmental protection and social cohesion, is not the right recipe for a happy future of humankind (Braun 1995).

Lean production and re-engineering

The technology manager is often called upon to manage manufacturing innovation, i.e. introducing new manufacturing machinery, control equipment, and processes bought from outside which are new in the context of the firm. The very act of choosing, and the act of getting the best out of purchased equipment, requires major effort. If the plant consists of many individual machines and devices, then welding these together into an efficient manufacturing organization requires months of designing, fine tuning, bug-hunting, training, negotiating and organizing (Bessant 1991).

Much manufacturing innovation in recent years has aimed at converting the manufacturing process into one of lean production. As the name implies, this means cutting out all unnecessary layers of 'fat', such as unnecessary tasks, labour, materials, stocks, rejects, hierarchical layers. The chief ingredients of lean production are design of products for ease of manufacture; layout of the factory for minimum transportation; reduction of stock of parts and materials by insisting that suppliers deliver 'just in time'. Just in time also means that parts arrive at the assembly line just when they are required and only minimal buffer stocks are kept. This eliminates the need for storage space and for capital tied up in stores. Lean production also means 'total quality control', so that all parts are within their tolerances and no mistakes are made in assembly, reducing the amount of costly re-working virtually to zero. The number of workers is reduced and they are multi-skilled, so that they can be deployed wherever they are needed. The hierarchy is flattened by cutting out all unnecessary layers and shortening the chains of command (see Womack *et al.* 1990, Bessant 1991).

In the recent past, the trend was to automate and computer control as much as possible, with the ideal seen as a factory with virtually no workers. This trend appears to have been reversed now, with the ideal seen as automation of repetitive and strenuous tasks, but with a workforce that is skilled, loyal and actively involved in increasing the efficiency of production. Too much automation may not be cost-effective, and it almost certainly contributes to unemployment and de-skilling. An enlightened manufacturer strikes a careful balance between automation of routine and strenuous operations, and the use of skilled labour where the flexibility, ingenuity, loyalty and creativity of people are major assets.

Process innovation, whether it consists of true innovation or of the installation of new machinery and processes bought from outside (manufacturing innovation) is crucial to the competitive position of the firm. Manufacturing technology is a major determinant of both the quality and the price of the product. It also determines the product mix that can be produced. The current emphasis in production processes is on flexibility, which allows the production of a mix of similar, but different, products and makes both short production runs and great variety of product an economical proposition.

The proper functioning of a complex system needs constant managerial attention and vigilance: routine maintenance must be carried out; malfunctions must be detected and rectified; there must be a constant search for, and elimination of, weak links in the complex chain of operations. If flexibility is to be attained and orders fulfilled on time, scheduling of operations becomes crucially important.

The administrative equivalent of the trend toward lean production is so-called re-engineering of information handling. All unnecessary administrative linkages and tasks are eliminated. The availability of powerful software packages makes it possible to streamline administrative tasks to a very considerable degree. Decisions that depend in specific ways upon pre-determined values of a number of variables can now be made by computer, thus removing the need for human discretion. In loan applications or insurance underwriting, for example, variables such as the applicant's income, family status, age, medical history, occupation, etc. can all be given ranges of values that affect the outcome in a pre-determined way. This removes human fallibility, but also human flexibility, understanding and responsibility. Tasks that previously required the attention of several skilled human operators, can now be dealt with by a very much reduced workforce. Only exceptional cases require human attention, routine cases are dealt with by computers operated by very few people. This trend may lead to the resolution of what was previously regarded as a paradox, when large amounts of investment in computers did not appear to cause any increase in the administrative efficiency of organizations. On the other hand, the new trend does not augur well for employment of white collar workers (Head 1996). Does it augur any better for the customers of service organizations?

2

PRINCIPLES OF
TECHNOLOGY ASSESSMENT

What is technology assessment?

If we take the words technology assessment in their ordinary everyday sense, they mean evaluating a particular technology, or technology in general, in some way. This interpretation forces us to think of the purpose of evaluation, for to evaluate means applying some kind of measure, and the measure we apply depends on the purpose for which we are evaluating. To evaluate technology in general is a task for philosophers, and we shall avoid it here, concentrating on the difficult enough issue of evaluating specific technologies.

Technology has become one of the principal weapons in the competitive struggle between firms. It can be used both tactically and strategically. How well a firm performs depends to a considerable extent on how well it understands, masters and uses technology. 'The competitive success of today's business clearly depends on the use of technology. Modern organizations face the dual challenge of keeping up with rapidly changing technology and making sense of it' (White 1988, 10).

To gain full competitive advantage from technology it is not enough to view its deployment as a narrow specialist task, where it is sufficient to ask whether the technology will perform a task at reasonable cost. We need to know a great deal more and technology assessment is the tool, or the frame of mind, that allows firms to examine technologies in depth and with foresight in the context of the firm's interests and capabilities, as well as in the context of the society the firm lives in. The objective of technology assessment is to consider technology in its full context, with all its opportunities, possibilities and ramifications for the firm and the environment in which it operates.

Consider, for a start, the apparently trivial problem of determining the utility of a certain technology for a specific purpose. We may, for example, ask whether arc welding is the best method for fabricating some particular metal structure. Evaluating a technology for its utility means comparing it with other technologies, for we do not only ask whether a particular technology will do the job, but also whether it will do it better, cheaper, or more easily than an alternative technology. On the face of it, answering these questions should suffice to evaluate a technology, but we also need to ask how a particular technology will fit into the firm with its given organization and skills. We need to ask what impacts the technology might have and what side-effects it might cause, whether on the product, the workforce, the factory environment, or on the wider environment. Evaluating brings into play questions of values and of both technical and social purpose: do we need the technology to give the best possible quality, or good value for money (to be defined from case to case), or should it present the least possible hazards to human health, or offer the best possible protection of the environment, or give maximum employment, or use maximum skills, or reduce the payroll to a minimum?

Leaving the trivial example and turning our attention to the more general case of evaluating technology for a specific technical purpose, we find that the apparently simple task of selecting a technology on its technical merit turns out to be fraught with difficulties of both a technical and a non-technical nature. The technical selection alone depends both on the chosen selection criteria and on subjective judgement. It needs judgement and knowledge to guess which technology is obsolescent and how soon it will be obsolete. It needs foresight and knowledge to judge how good the new alternative will be. Decisions need to be made on which firm is likely to supply the best of the chosen alternative, or whether the chosen technology should be developed and produced in-house. It needs knowledge to foresee all the organizational and personnel consequences that different technical solutions might bring in their wake.

Selecting a technology on non-technical criteria is even more difficult. First, the setting of criteria becomes dominated by political judgement: judgement on what is best for the firm in the long run, and judgement about future legal requirements and public preferences. Second, the analysis of the environmental and social consequences of technological choices is difficult and uncertain. Because technological development often proceeds in small steps, and each such step has negligible social and environmental consequences, it is very hard to foretell where the sum of such steps will lead and what the integrated effects will be. It is the art and science of foreseeing the effects of technological change that lies at the core of technology assessment. Even more important, however, is the attitude of mind which attempts to take a broad and far-sighted view of introducing a new technology. Technology assessment demands not only that we should look before we leap, but that we should look beyond the obvious, that our horizon in time and space should stretch as far as is humanly possible.

We define technology assessment – in the context of technology management – as *a systematic attempt to foresee the consequences of introducing a particular technology in all spheres it is likely to interact with*. The essential meaning of technology assessment is that making technological choices should be preceded by a thorough analysis of all the consequences the choice might have, not just the immediate and sought-for consequences. Technologies often have had unintended effects which, on occasion, have proved highly undesirable. Some of these might have been foreseen and avoided if an attempt to look for them had been made.

The modern concept of technology assessment (TA) arose in the public domain. It was intended as a source of information and as an input into the policy making process. The formation of public policy toward technology, whether for the support or for the control of technology, is difficult and important and requires the most comprehensive and far-sighted information that can be made available.

The private domain is no different. What policy a firm chooses, how it goes about obtaining the right technology, how it applies it, how it renews it; all these are vital questions that play a major role in determining the firm's fate. If the management of technology – the private equivalent of public policy formation – is to be successful, it too needs the best available information. The bulk of this book deals with technology assessment in the commercial domain, but to understand the origins and principles of technology assessment we need to look at its beginnings in the US Congress.

Technology assessment in the United States Congress

The beginnings of technology assessment in the modern sense can be traced to the United States Congress. During the 1950s and 1960s, more and more demands were made on Congress with regard to financial and other support for technological innovation, and for legislation controlling the undesirable effects of technology.

Demands were thus made for the formulation of both aspects of technology policy: support for technology and control of technology. One of the best-known examples of a request for financial support was that for the construction of a civilian supersonic airliner. At a time when Britain and France were pressing ahead with the design and construction of Concorde, there were many voices in the US clamouring for a bigger and better supersonic airliner. It took till 1970 for Congress to reach the decision not to support the project, thus causing it to be abandoned. No commercial firm would bear the enormous costs and risks of such an enterprise. The battle for and against the US supersonic airliner was bitterly fought and many members of Congress and its committees felt the need for more comprehensive and more objective information, to supplement the special pleading by supporters and foes of the project.

Demands for the control of technology were made when concerns were voiced about the dangers of lead in paint, especially paint used for children's cots. Further concerns began to emerge about the use of persistent pesticides and eventually DDT was banned. Similar concerns were raised about the use of chemical fertilizers and industrial chemicals, which reached the water courses and led to the death by eutrophication of some rivers and lakes. All these issues are complex and controversial. The chains of cause and effect are hard to establish and the economic interests involved are very substantial. Members of Congress mostly lack training in science and find the issues confusing; though even trained scientists find it difficult to reach conclusions on matters where cause and effect can only be surmised and where hard facts rarely form an adequate basis for a policy decision.

The immediate post-war period had come to an end and with it the simple belief that technology was a 'good thing' that helped to create wealth and was able to solve most problems. Doubts about the previously unquestioned benefits of science and technology had emerged and much attention was drawn to unwanted, undesirable and dangerous side-effects of technology. Disillusionment with science had begun and demand for control policies was growing. At the same time, demand for support of technology was also growing, as technological innovation had become more expensive and more risky. The result of all this was that Congress was faced with an increasing work-load on issues related to technology and found itself floundering in a morass of pleading and counter-pleading. What it wanted was objective reports setting out concisely and accurately what was known about the technologies in question and about their potential impacts. What Congress also wanted was an information gathering service under its own control, rather than under the control of the administration. It was Congressman Emilio Daddario, chairman of the House Committee on Technology in the late 1960s, who was instrumental in formulating the need for technology assessment and for establishing the Office of Technology Assessment (OTA) of the US Congress in 1972. The bill setting up the OTA states, inter alia,

> It is, therefore, imperative that the Congress equip itself with new and effective means for securing competent, unbiased information concerning the effects, physical, economic, social, and political, of the applications of technology, and that such information be utilized whenever appropriate as one element in the legislative assessment of matters pending before the Congress.
>
> (from a 'Bill to establish an Office of Technology Assessment', put before Congress on 19 July 1971; reproduced as an appendix in Medford 1973)

One might be tempted to believe that issues of technology policy can be resolved by marshalling all the facts and making a decision based on the balance of factual argument. In reality, however, there can be no factual information about the future. Often there is insufficient knowledge about hazards and impacts of technologies. The assessment of how great a risk is, and whether it is worth taking, must always remain a matter of opinion. For example, the issue of the safety of nuclear electricity generation and the final storage of spent radioactive material is as controversial today as it ever was, and that after some fifty years of intensive research and debate. Thus the task of TA goes beyond marshalling facts or even casting tentative glimpses into the future; it must try to distinguish between fact and conjecture and review arguments on controversial issues.

During the several years of discussion that preceded the establishment of the OTA and in the years that followed, many alternative definitions of TA were suggested. Only the one given by the Congressional Research Service of the Library of Congress shall be quoted:

Technology Assessment is the process of taking a purposeful look at the consequences of technological change. It includes the primary cost benefit of short term localized market economics, but particularly goes beyond these to identify affected parties and *unanticipated* impacts in as broad and long range fashion as is possible. It is neutral and objective, seeking to enrich the information for management decisions. Both 'good' and 'bad' side effects are investigated since a missed opportunity for benefit may be detrimental to society just as is an unexpected hazard.

(Hetman 1973, 57)

What Congress required was interdisciplinary, future-oriented advice on the full spectrum of consequences of the application of new technologies, and guidance on technological options. Congress sought more complete, more reliable, less biased and less voluminous information than would otherwise be available. The 'Founding Fathers' of Technology Assessment imagined and believed that by equipping legislators with an objective analysis of the foreseeable and less foreseeable effects that might result from the application of a new technology, it would become easier both to control and to support technology in ways which would provide optimum utility, and cause least damage to society.

The early definitions of TA should be viewed as statements of an ideal. It is obviously beyond human capabilities to achieve objective, all-embracing descriptions of the future impacts of a technology, let alone the effects of a miscellany of possible combinations of technological and socio-economic developments. Objectivity, foresight, omniscience, all these are attributes which the ideal technology assessor should, and cannot, possess. Ideals are useful, however, as beacons guiding us in the desired direction, even if we know that the ultimate actual achievement must fall short of the goal. The quality of a technology assessment must be judged by the closeness of its approximation to the ideal, by the reliability of its information, and by its humility in acknowledging uncertainties and gaps in knowledge and understanding. Technology assessments can never be perfect, but they can be of high quality and of high utility.

Generally speaking, assessments are carried out by small teams of expert assessors, aided by advice from experts in the relevant fields of knowledge and seeking inputs from relevant interested parties (stake-holders). Clearly, perfect objectivity is not given to humans who have beliefs and adhere to systems of values. Nevertheless, a team of assessors who attempt to be as objective and as truthful as possible, and who collect information from a full range of available sources, should provide a reasonable approximation to objectivity. One of the conditions that need to be fulfilled, however, is that the team must be free of fear; its members must be willing and able to say what they see as the truth, not what they guess their bosses want to hear. The ideal assessor should be of an inquisitive independent mind and of a sceptical disposition – weighing up all evidence carefully to obtain a true balance.

The requirement of illuminating a problem from all sides needs to be modified in practice both for financial reasons and because success in science is born from the correct selection of relevant aspects, not from getting hopelessly lost in a welter of scarcely relevant detail. The true art consists of knowing what is, or will become, important. There are diminishing returns on investing too much effort into too much detail. The selection of features of a problem for detailed examination is part and parcel of the study and requires insight, iteration, discussion and judgement.

The ultimate stumbling block on the path to achieving a perfect technology assessment is the need to look into the future. Technology assessment with hindsight can teach us a lot about fundamentals of the relationship between technology and society; but only forward looking TA can be of any utility to decision makers. Decisions are concerned with the future, not the past. Technology assessments invariably contain several elements of forecasting: technological forecasting for the technology under discussion, as well as for its rivals and complements; social, economic and political forecasting in an attempt to foresee what sort of world the technology will be interacting with; and, finally, forecasting the wide range of varied effects, obvious and less obvious, that the technology might cause.

To say that forecasting is prone to error is not only a commonplace, but also an understatement. Nonetheless, it is an essential aspect of the human condition that we attempt to foresee the future in an effort to shape it. Forecasting must be viewed as a tool for moulding the future and as a dynamic, interactive process. Forecasts do not assert what the future shall be; rather they attempt to glean what it might be and what opportunities for shaping it might offer themselves. Technology assessment uses a range of interrelated forecasts to view the future technology in a future setting. It goes beyond this to suggest possible policy measures that could alter the future in ways desirable to the decision maker. The assessor suggests a range of possible policy options; the decision maker decides which of these, if any, to implement.

The fact that TA looks into the future is one of the reasons why technology assessments have to be repeated at intervals as the future unfolds. There are two further facets to TA which underline the fact that assessments have a limited life and have to be up-dated whenever the need for new decisions arises in a substantially changed situation. First, science and technology change very rapidly. Second, for TA to be useful it must obtain information from wherever it resides, and much of this information consists of opinions, attitudes, fears, interests and hopes, and is thus subject to change and shifting of positions. Indeed the very process of conducting a technology assessment involves much discussion and dialogue and thus contributes to the formation of opinion. It has been said that if all the information contained in a TA report has not diffused by the time the report is published, then the process of conducting the technology assessment was faulty.

We can gain further insight into the nature of technology assessment by looking at the basic methodology of conducting technology assessments. This was defined in the very early days of TA in the United States and has not changed since. Although TA and its methodology were introduced as an aid to public policy formation, both the principles and the methods bear much resemblance to TA conducted as an aid to management decisions in commercial firms. Though there are differences in what is required of technology assessment in the public and in the commercial domains, the basic methodology is the same in all cases.

Basic methodology for TA

The first step

The first issue to be resolved before embarking upon a technology assessment is the topic to be treated. The topic is usually a technology, but it may be a social problem that might be ameliorated by the application of technology. We speak of technology oriented TA or of problem oriented TA. The process of TA itself is virtually the same in the two cases and we shall treat them mostly without drawing the distinction. In the case of public policy, the technology or problem to be assessed must be of some interest and concern to the legislators. It must thus be a technology that is on the political agenda either because it might require public support or because it might have to be controlled by legislation, or both. For preference, the issues concerning the technology should not be burning ones; TA is a strategic tool rather than an aid to fire-fighting. The above remarks need only very slight modification to be applicable to managerial problems in commercial firms. Indeed, public and private concerns are often closely related.

The next issue to be resolved is the scope of the assessment. Is it to look at only one narrowly defined technology and its immediate rivals, or is it to look at a major bundle of technologies serving related purposes? For example, is an assessment to be made of, say, glass fibre optics, or of information technology? In another example, is the TA to look into the future of hydrogen as a fuel for motor cars, or is it to be concerned with the future of road transport? Choose too narrowly, and the assessment will be unlikely to reveal anything of great interest; choose too widely, and the assessment will not only become very expensive, but it will also be unwieldy and unlikely to prove of much practical help to the decision maker. When deciding on the scope of the topic, we also need to decide on the time horizon to be covered. A long time horizon brings with it very great uncertainties; a short horizon may be insufficient to reveal truly important aspects of the problem.

Decisions about topic, scope and time horizon are extremely important and must be taken in consultation between assessor and decision maker. Indeed, the determination of the correct scope should form the first part of the assessment. We call this the first step, though perhaps it should be termed the zeroth step because it is preliminary to the assessment proper. The first (or zeroth) question to be answered is thus: what is to be the topic, the scope and the time horizon of the technology assessment to be undertaken?

The further steps to be described are logical ones and not necessarily a temporal sequence.

The second step

The second step of any TA consists of a description of the technology under scrutiny, or of the technologies relevant to the solution of a problem under discussion. In addition to merely describing the main technology, a description of alternative, complementary and rival technologies must be included and some considerable thought ought to be given to likely development paths of all the technologies described. Complementary technologies are those that are needed to make a technology feasible and practical, or more effective and wider in scope. For example, the development of integrated circuits depends critically upon technologies such as growing single crystals of silicon of very high purity, on clean room technology, on photo-lithography. The scope of telecommunications is extended by the use of answering machines, memories, telecopiers, and so forth. Rival and alternative technologies, on the other hand, fulfil much the same function as the technology under discussion. It would be foolish, for example, to discuss satellite TV broadcasting technology without mentioning the rivalry between it and cable television. Alternative technologies are those that can substitute for the main technology. Particularly in process technologies it is often possible to achieve the same result in several different ways. The terms alternative and rival have much the same meaning, though rival puts more emphasis on the direct competition between the technologies.

Performance, principles of operation, as well as costs are regarded as major descriptors of technologies, though forecasting the latter often proves misleading in the extreme. The technology descriptions should be couched in a language accessible to the intelligent lay-person and thus be truly informative for the politician, the civil servant or the general manager.

This part of the TA goes beyond being an exercise in scientific journalism. It contains an element of analysis in that it asks questions about related technologies and the technological system, and it contains an element of forecasting in that it attempts to foresee the future of the technology under scrutiny, as well as that of its rivals and complements.

Put at its simplest, the second step of a technology assessment provides an answer to the question: what is the technology we are talking about, how does it fit into the technological system and how is it likely to develop?

The third step

The third step of a technology assessment concentrates on the core questions: what benefits are to be expected from the technology, what needs does it satisfy, and why is it superior to present, or rival, technologies? The benefits may be purely economic and commercial, or they may be expressible in terms of environmental improvement, health benefits or, on a more subtle level, improvements in the social or political fabric of society. By speaking of benefits we do, in a sense, apply a value judgement, though there is pretty universal agreement about some of the 'good things' that technology has to offer. In any case, the analysis will attempt to describe the non-controversial benefits as well as those impacts that might be regarded as benefits by some and not by others. It will also attempt to show who the benefits, such as they are, might accrue to and will not shun pointing out controversies.

In general, questions of utility are laden with value. On the assumption that in a capitalist and liberal society the willingness to purchase is a sufficient yardstick for utility, commercially viable technologies are deemed to be useful, unless proved otherwise by showing that they cause some harmful effects. On the other hand, some technologies may offer considerable benefits to society, yet are commercially too risky or otherwise too unattractive to become viable without some form of public support. The amelioration of environmental damage, for example, is entirely in the public interest, yet no viable market mechanism has, in general, been found to ensure the funding and advance of environmentally beneficial technologies without public support. The final arbiter of worthiness for support must be the political decision maker, although technology assessment can carefully marshal and weigh the arguments advanced both in favour and against the beneficial effects of the technology and its dependence on public magnanimity for survival.

The basic question that the third step of a technology assessment attempts to answer is: what benefits can be expected from the technology and to whom will they accrue?

The fourth step

In the fourth step a technology assessment must address the question of what unwanted effects or hazards the technology might cause. When such undesirable impacts or dangers are identified, the associated problem of who or what might be adversely affected needs to be addressed. In close analogy with the previous point on beneficial effects of a technology, it may be a matter of controversy to decide whether an effect is 'good' or 'bad'. In any case, close attention must be given to all possible side effects, which might prove destructive of the natural environment, dangerous to human health, disruptive of society, or otherwise trigger a chain of events which appears unpredictable and risky. As far as possible the risks and impacts ought to be described in quantitative terms, though very often these are artificial and meaningless and it is wiser to stick to qualitative statements. Generally, only the measurable ought to be measured; for the immeasurable, verbal or graphical analysis must not only suffice, but is indeed superior. Often hazards associated with the use of technologies are mere hunches and it is an important task of TA to collect the evidence and show consensus or controversy. It needs to be emphasized that the answers TA can provide cannot be better than those provided by the individual disciplines concerned, except that the whole of a TA report ought to give better insight than the sum of its parts. If climatology does not know for sure what carbon dioxide in the atmosphere will do to the climate, then neither does TA. If economists and sociologists cannot tell whether microelectronics will cause unemployment or de-skilling, then neither does TA. Ignorance does not turn into knowledge, uncertainty into certainty, by merely re-labelling the package. What can be expected of TA is a well-documented, complete and honest assessment of the state of the art; neither more nor less.

The simple formulation of the question addressed in the fourth step of a technology assessment is: what dangers does the technology harbour and what ill-effects might it cause?

In both the third and fourth steps of the analysis, i.e. those steps that seek to describe the effects, or impacts, of the technology, the analysis must go well beyond the obvious. It is not only first order effects, but also higher order effects of effects, that we are seeking. As it is frequently difficult to distinguish a priori between positive and negative impacts, steps three and four are often combined into a single impact assessment.

The fifth step

The fifth and final step in a technology assessment consists of an analysis of policy options. If a technology may require supportive measures for its development and diffusion, the analysis should show why this is likely to be the case and what measures might be available. The technology may not require any intervention and may thrive happily in a purely commercial domain; on the other hand, it may need help in terms of grants, tax allowances, training or information programmes, legislative or regulatory measures, or administrative and institutional arrangements. In the case of a technology assessment carried out for a commercial firm, the policy measures will be rather different. It is then not a matter of public policy (although the firm might be able to take advantage of public policy support, or might have to take action to comply with control measures) but a matter of management policy.

An analysis of possible measures for the control of the unwanted effects and risks, and for the likely efficacy and efficiency of such measures, must form an integral part of any technology assessment. As has been stated before, benefits and risks will be described as such only in cases of widespread consensus. In the absence of consensus, it is for the decision maker to decide whether an expected effect is desirable or otherwise.

In this context it is necessary to remark once again upon the respective roles of politics and of TA. The TA analyst provides descriptions of costs and benefits, attempts to quantify them, and assigns both costs and benefits to affected parties. He or she also describes perceived difficulties in the path of the technology and provides an analysis of available options for supportive action. Similarly, unwanted effects and the affected parties are identified, and an analysis of available policy options for the control and amelioration of the dangers is provided. Each policy option needs to be analysed in terms of its effect upon the impact of the technology. How does it ameliorate unwanted effects and how does it enhance benefits?

The analysts try to be as objective as possible and to present as balanced a view as they can manage. They do not offer advice of the kind 'this ought to be done', but only of the kind 'if you wish to achieve this and that, the following instruments are at your disposal and may prove effective'. The strength of TA lies in its humility, in knowing its preserve and not usurping a role which does not legitimately belong to it. It has often been argued that scientists do, in effect, provide advice in their treatment of issues, in ignoring some facets of a problem and stressing others. All this may be so, but this should merely be the difference between human frailty and ideal achievement, not a deliberate act of using the disguise of TA to become a back-room politician.

In summary, the fifth step of a technology assessment answers the question: what support or control might be needed for the technology and what options are available for providing them?

The basic methodology may be summed up as: scope, technology, impacts, policy (STIP).

When we attempt to foresee the fit of a technology into society at some future date, we usually employ a method of scenarios. That means that we make several different assumptions about the way both the technology and society will develop and describe the futures that we might expect if the different sets of assumptions were to correspond to reality. The assumptions we make usually cover aspects such as high economic growth or low growth; attitudes of concern over the environment or attitudes of lack of concern; high or low energy consumption; and such like. In other words, we consider a probable range of future values for a few social variables, including dominant policies, that might affect the use and impact of a technology. Scenarios essentially consider a range of probable effects, occurring within a range of probable boundary conditions.

In the early days of technology assessment much effort was put into developing methodologies. It was thought that TA should become a new scientific discipline that, to be respectable, needed its own methods of enquiry and analysis. The early obsession with methodology soon gave way to pragmatism. 'Technology assessment is merely a recommended method, not a scientific discipline. . . . It is a logical approach that requires analytical skills' (White 1988, 13). It was found that each assessment required a somewhat different approach – horses for courses – and that a general TA methodology as described above, coupled with methods borrowed from a variety of disciplines, were all that was required. Early writings on TA discuss the problems encountered in attempting to obtain the greatest possible social benefit from technology, but they also describe in detail attempts to develop suitable methods of analysis (Hetman 1973; Medford 1973). Later writings are less concerned with methodology and more concerned with making TA an effective aid to policy makers. Keller (1992) discusses the difficulties of providing policy analysis in a Congress dominated by politics, and regards the OTA as an attempt to resolve the difficulty. As we know now, the attempt failed in the long run, though it was quite successful in the short run.

We shall return to methods in Chapter 5, but for the sake of slightly more completeness and historical accuracy, we shall briefly describe the most commonly quoted original general methodology for TA (Hetman 1973, 119). The methodology consists of seven steps:

1　Define the assessment task: discuss relevant issues and any major problems; establish scope (breadth and depth) of enquiry; develop project ground rules;

2　Describe relevant technologies: describe major technology being assessed; describe other technologies supporting the major technology; describe technologies competitive to the major and supporting technologies;

3 Develop state-of-society assumptions: identify and describe major non-technological factors influencing the application of the relevant technologies;

4 Identify impact areas: ascertain those societal characteristics that will be most influenced by the application of the assessed technology;

5 Make preliminary impact analysis: trace and integrate the process by which the assessed technology makes its societal influence felt;

6 Identify possible action options: develop and analyse various programmes for obtaining maximum public advantage from the assessed technologies;

7 Complete impact analysis: analyse the degree to which each option would alter the specific societal impacts of the assessed technology discussed in step 5.

The five steps described in our methodology cover virtually the same ground as the seven steps described here. The only difference is that we specifically mention hazards caused by technology and policy measures for their control, whereas the above seven steps speak only of impacts and of policy measures for obtaining maximum public advantage from the assessed technologies. However, impacts can be both positive and negative, and maximum benefit includes the avoidance of ill-effects. Developing state-of-society assumptions is much the same as considering prevailing socio-economic conditions and developing scenarios. The seventh step is part of the fifth step of our five-step methodology. Policy options clearly need to be examined for their likely effects upon the impacts of the assessed technology.

Let us recapitulate the list of questions, simplified to the point of naivety, which are within the competence of TA to answer: *What is the technology we are talking about, how does it relate to competing and complementary technologies, and how is it likely to develop? What benefits does the technology bring with it, who will benefit from it, and what help does it need to be given? What dangers does the technology harbour and what can be done to control them?*

In its early days, TA was intended to serve exclusively the infor-mation needs of the US Congress. It was TA in the political arena and it stood or fell on how well it served the needs of political deci-sion makers. TA can serve the decision maker if, and only if, certain conditions can be met – if some form of matching between the requirements of the political system and the capabilities of the technology assessment system can be arranged.

The same is true, of course, when we speak of technology assessment undertaken for the benefit of managerial decisions. In the industrial context, as much as in the public policy context, TA can thrive only if it serves the needs of decision makers. There are very close similarities between public and private management of technology. Substitute the word manager for the word politician, imagine a decision maker in a commercial firm rather than a decision maker in the public domain, and the similarities between their roles in decisions about technology appear to be very marked.

The fit of TA into politics

The first condition for the effective use of technology assessment in politics is that TA should be concerned with matters which form part of the political agenda. If TA deals with matters politics does not deal with, its results will be ignored. If TA is too late in providing answers to political problems, the problems will be solved without the benefit of TA. This means that TA must be given sufficient warning of impending political agendas to be able to prepare its information, and there must be effective mechanisms which enable TA institutions to undertake the required studies. These conditions were met in the case of the OTA, but are not automatically met in different institutional settings. In the industrial context: 'Organizational receptiveness to technical change is a key issue that cannot be avoided by the assessment consultant' (White 1988, 43).

The setting of the agenda for technology policy issues is a complex process. The participants in this process are the politicians and their grass-roots informants; industrialists and their lobbies; scientists and their pressure groups; citizens' groups; and the media. Technology assessment institutions have a role to play in that they have contacts with many sources of information and are accustomed to look into the future in matters technological. In the early days it was thought that technology assessment institutions could act as a kind of watchdog to alert politicians to dangers looming on the technological horizon. To some extent, this remains true, though the political emphasis is now much more on spotting hopeful and promising new technologies, rather than on spotting dangers arising out of new technologies and new knowledge about older technologies. Even in

41

cases when it has been decided that a technology requires political attention, the question of whether or not the subject is suitable for the 'technology assessment treatment' needs to be decided in collusion between technology assessors and politicians. Some technological issues may be simple enough to be resolved without much ado, some may be too urgent to be able to wait for the long-winded process of a full technology assessment, though some form of mini- or micro-assessment may be suitable for them.

The second condition which must be met is that a relationship of trust should exist between the decision maker and the technology assessor. Indeed, only if such trust exists will the assessor be given the early warning, the time, and the money needed to carry out the assessment. Only if trust exists will the politician listen to the assessor and take the analysis seriously into consideration when reaching decisions. In order to establish trust, the assessor must have an impeccable scientific pedigree, must be seen to be independent, must take great care to fulfil the conditions of impartiality, must listen carefully to advice and opinions, and must keep the decision maker and all interested parties fully informed throughout the process of assessment.

It is vital that the final report produced in a technology assessment, as well as its various abstracts and derivatives, should be readable, brief and informative. A comprehensive executive summary is of the essence. The busy executive, who only reads the summary, need not know all the detail, but must be aware of all essential points of the report. Lengthy abstruse discourses, written in scientific jargon, are of no use to man or beast in the practical world.

The politician, aided by the normal processes of political consultation and political advisory and adversary mechanisms, must decide which, if any, of the many possible actions suggested in a technology assessment should be taken. This is an act of political will, driven by political forces, but aided by the information gathered through a technology assessment. Only the political process can decide how the suggested policies fit into the general political framework, whether they are likely to prove acceptable to the public, and how internal conflicts of interests or policies can be resolved.

These matters are important to the student of technology management. First, because of the close similarity between the roles of technology assessment in public policy formation and in industrial management decisions. Second, the technology manager needs to understand how public policy toward technology is formed, because the commercial enterprise operates within a framework set by the state and its policies. Public technology policy may play a crucial role in the life of a technology and, even more important, in the life of commercial firms involved in the production and use of technology.

The Office of Technology Assessment thrived for twenty-three years and produced 755 technology assessment reports of one kind or another during that period. Unhappily, it was closed down in September 1995, though not without having fathered a number of imitators in many countries. The official reason for the closure of the OTA was budgetary; the real reasons can only be surmised. Some believe that the OTA was the sacrificial lamb offered by Congress on the altar of budgetary restraint. It seems, however, that the spirit of technology assessment in the political arena is coupled with a spirit of belief in the possibility of positive action by governance. The spirit of TA is allied to a spirit of open enquiry, a spirit in which politicians believe that they have an active positive contribution to make to the welfare of the commonwealth. Politicians who believe that their only proper contribution is to trim public spending, to retreat from state intervention wherever possible, and that the only truly legitimate role for the state is internal and external defence – the police and the armed forces – see little need for technology policy. Surely technology can be left to private firms and any ill effects that come to light will eventually be eradicated by market forces. And the natural environment? Just scaremongering. Technology assessment in the public domain is not patronized by politicians who hold these beliefs.

There can be little doubt that the OTA served a very useful function during its life, and its many imitators are a kind of testimony to its utility. There can also be little doubt that technology assessment in commercial firms is here to stay, for the spirit of laissez faire is the last thing managers want in their enterprises. Managers firmly believe in good management, in forward planning, and in not leaving technology either to chance or to the unbridled whims of technologists and scientists. Technology assessment is also here to stay in the service of many central and local governments. A study of the Institute for Technology Assessment of the Austrian Academy of Sciences lists six TA institutions serving the needs of parliaments in Europe (Peissl and Torgersen 1996, 33–49). There are many more TA institutions with mixed or different missions throughout the world. To quote a recent example, the local governments of Greater Vancouver have embarked upon a programme of improving living conditions over the next 25 years by reducing dependence upon the motor car, reducing encroachment upon green land, and so forth. The plan was adopted after extensive research that has all the hallmarks of a technology assessment (N. R. Peirce, *Guardian Weekly*, 25 August 1996, 17).

The fields of activity of the OTA covered the full range of public concern with matters technological: Defence, Space, Energy, Environment, Education, Transport, Health, Economy, Materials, Telecommunications. The OTA involved many advisers and listened to much argument by stake-holders. Workshops with stake-holders were a regular feature of their information gathering process. This helped the OTA to be well informed, but it is also an important step toward achieving objectivity. Most observers regard the OTA reports as authoritative and reliable. The source of the OTA's greatest pride was the fact that in debates in Congress and its committees, members on all sides and of all shades of opinion used arguments and information from OTA reports. This appeared to be the ultimate hallmark of utility and, more important, of true objectivity.

Types of industrial technology assessments

There are many different types of technology assessment, depending upon the technologies or problems to be assessed and on the ground to be covered, i.e. the scope of the assessment. Harvey Brooks (1992) distinguishes five types of TA: project assessment; generic technology assessment; problem assessment; policy assessment; global problematique. Except for the last category, the words are self-explanatory. Global problematique means analysing global issues such as climate change or, slightly less global, an issue such as the role of modern bio-technology in development (United Nations 1992). Brooks further distinguishes eight categories of scope of assessments. Rather than go into details of these categories, let it suffice to say that not all assessments look into all possible – or even all necessary – aspects of the problem in hand. Many assessments are only partial and fall short of the all-embracing ideal. This is not to say that the scope of assessments does not need to be limited, as otherwise there is a danger of getting lost in an infinity of aspects that just might be relevant. The scope needs to be chosen wisely, to include everything that is essential and exclude all unnecessary detail. Sometimes only partial assessments are carried out and these may be useful, as long as everybody is clear that they are partial[1] and what their scope is.

In the industrial setting, technology assessment 'seeks opportunities to match changes in techniques, processes, and equipment to specific business goals and objectives' (White 1988, 40). Technology assessments are likely to be required in the following situations, each of them representing a type of assessment.

1　A new production (process) technology comes, or is about to come, onto the market. If it appears at all likely that the new technology might be of interest to the firm, its potential, together with all the consequences of its introduction, costs and advantages, needs to be assessed. Ways of acquiring and introducing the new technology need also be considered. Alternatively, a weakness may be discovered in the production processes of the firm. Ways and means of remedying this weakness, together with the implications of each possible remedy, need to be examined.

This situation calls for a *process technology assessment*. This means that the main emphasis of the assessment is to examine the new process technology and its alternatives, and looking at all the impacts these technologies might have on the firm and its surroundings.

2　A decline in the industry the firm operates in is becoming apparent. An assessment needs to examine whether the decline is inevitable, what defensive action is possible, or what other means of escaping decline and oblivion might be available.

This situation calls for an *industry technology assessment*. Thus the main emphasis is on examining the whole branch of industry and looking into the causes of its decline. The TA also needs to examine possible political and technological remedies and analyse the full range of policy options.

3　A particular product market is declining. The reasons for this need to be examined and means of reviving the market need to be found. The decline may be caused by shifts in consumer behaviour, by the obsolescence of the product, by the wrong price/quality relation, by superior performance of rival firms, or by general economic conditions. All the consequences of introducing an alternative product, alternative production methods,

---

1　In truth, of course, any assessment carried out by humans is incomplete, but some are more incomplete than others.

or alternative markets need to be examined. Alternatively, a major new product idea is developed, either within the firm or on the market. The idea for the product may be a result of scientific research or it may be a response to new socio-economic conditions, or to new environmental concerns or regulations. The potential for the new product needs to be examined, together with all the adjustments and investments that need to be made within the firm if this product is to be manufactured and marketed.

The type of assessment called for by the above situations is a *product technology assessment*, i.e. a full technology assessment with greatest emphasis on a particular new product and alternatives to it.

4   A new political constellation is discernible, such as the European single market or new GATT international trade agreements. Alternatively, new regulations concerning safety at work, environmental protection, or product safety may be about to come into force. Such changes in political and regulatory conditions need to be examined. They offer opportunities for new products, or for overdue changes in production methods. Even if no advantage is to be gained, the new rules must be complied with and that may require changes with major repercussions.

Assessing the above situation may be regarded as a *political technology assessment*. This type of TA examines the full range of threats and opportunities arising out of the new situation. It analyses policy options and their consequences.

5   Finally, and very importantly, the firm needs to formulate its strategic plan. As we shall see below, technology plays a major role in strategy. A technology strategy for the firm, fitting into and forming part of the general strategy, needs to be formulated and analysed.

We call this a *strategic technology assessment*. Indeed, it is the task of formulating and examining strategic technology options that is generally regarded as the central task for the industrial technology assessor.

We thus distinguish five types of industrial technology assessment, though without claiming that these types are the only possible ones, or that they always need to occur in their pure form. Hybrid assessments are perfectly feasible. Though we gave a brief description of the main characteristics of each of the five types of assessment, they all normally follow the general STIP methodology.

We shall now describe a brief example of an industry technology assessment, taking the telecommunications industry as our example, mainly with a view to illustrating the general TA methodology.

Technology assessment of telecommunications

To illustrate the general methodology we shall look at an imaginary technology assessment in telecommunications. The example will not be fully worked out – that would exceed the framework of a brief text by far – but will be presented in outline only. We assume that it is carried out in about 1990 on behalf of a public telecommunications operator (PTT) who holds a near monopoly in a small European country (the client). The team of analysts for this TA consists of a telecommunications engineer, an economist, a policy analyst, a lawyer and a generalist team leader, well versed in TA. The team consults with stake-holders, such as the management of the client company, telephone users, officials of the ministry in charge of telecommunications, telephone equipment manufacturers, operators of mobile telephones and of cable and satellite television. A market research organization is used to make a forecast of various modes of telecommunications traffic. Members of an advisory committee, consisting of several stake-holders and further experts (a sociologist, engineers and economists), serve as advisers to the team and critically read early versions of the report. The final report consists of 300 pages of print and a thirty-page executive summary. The clients consent to making the executive summary widely available.

Following the steps of the general (STIP) methodology described above, modified to suit the particular case, the analysis proceeded as follows:

Step 1: scope of the assessment and time horizon

The technologies to be described and for which a development path is to be forecast are: telephone switching technologies; transmission technologies; domestic and commercial customer premises equipment (CPE); new modes of traffic additional to voice telephony. The assessment should look into technological and probable cost developments in these areas and at likely patterns of consumption.

The main impacts to be studied are: employment prospects in the telecommunications industry (operators and equipment manufacturers); changes in the terms of trade; the impact of regulatory developments (including European integration) on the client, on telecommunications users and on the general economy.

In view of very rapid technological and regulatory developments in the industry, the agreed time horizon for the study was to be ten years only.

Step 2: description of the technologies

The technologies to be described are all undergoing rapid development. The most striking thing about them, however, is that rival technologies are springing up for all traditional telephone equipment. Take transmission media. In the past, it was only the humble twisted pair of copper wires. Then there was the addition of microwave transmitters, then satellites were added and now the copper wire is being replaced by glass fibres that are rapidly coming down in price and can transmit infinitely more information per unit time (in telecommunications jargon: they have a greater bandwidth). There is another race on, however, because computer compression techniques make it possible to reduce the amount of information that actually needs to be transmitted, even though a lot of information is passed on. In a moving picture, for example, only those parts of the picture frame are transmitted that have actually changed since the last frame was sent, thus drastically reducing the amount of information that needs to be transmitted to obtain the full moving picture at the other end of the line. The mode of transmission, which previously was analogue, has mostly changed to digital. This makes it necessary to equip all exchanges (central offices) with digital equipment and makes the distinction between transmitting the spoken word and all other forms of information obsolete. Hence a lot of talk about ISDN (integrated services digital network).

On the subscriber's premises things are also changing rapidly. Whereas previously plain ordinary telephony (POT) was the only equipment in the private home, and business did not rise above a private branch exchange, now there is a very wide range of equipment available: facsimile machines, answering machines, computers linked to the telephone network, (electronic mail is developing, but the Internet is still nebulous), video-telephony, especially used for conferencing.

Many new services are being introduced or envisaged: home banking, home shopping, cash machines, tele-conferencing, distance consultancy (medical and technical, including, for example, the transmission of X-ray images), distance learning using moving images. Firms are developing sales networks, financial networks, parts ordering networks, networks for cooperation at a distance in computer aided design. Stock market information is being transmitted and banks use telecommunications to transfer funds. Mobile telephony is spreading rapidly, with a European digital system being developed.

Steps 3 and 4: impacts

Telecommunications technologies have no adverse physical effects. They do not degrade the environment or cause severe hazards to safety and health. The social and economic consequences, though considerable and varied, are hard to classify into positive and negative – it all depends on points of view. For these reasons the third and fourth steps have been combined into a single chapter analysing the impacts of the new technological developments on various aspects of society.

Total demand for telecommunications services is bound to increase, as more and more services are being offered and the technological developments are such as to drive costs (and, hence, prices) down. It has been argued that a fall in prices is caused by increased competition. Though this may be true, it should not be forgotten that the new technology alone – without increased competition – causes prices to fall. An electronic digital exchange costs less, performs better, and needs less maintenance than a mechanical equivalent. Investment in telecommunications equipment will have its usual multiplier effects, but as the production and maintenance of the new technology demands less labour than the old technology, the net employment purchased by each unit of investment on telecommunications equipment will decrease. Similarly, each unit of money spent on telecommunications services will provide less employment than in the past.

Many volumes have been written about the overall employment effects of information technology, including telecommunications. If tele-services become additional to traditional services, growth will result. If they merely replace traditional services, no growth will be caused. If the increased use of telecommunications improves the overall efficiency of the economy, this should lead to economic growth. Whether this growth will result in increased employment remains uncertain. Though it was expected that administrative services should become much more efficient, this effect was hard to measure. It is hard to measure efficiency in administrative services in principle, and, paradoxical as this may seem, it might take many years, if not for ever, for telecommunications and information technology to have a discernible effect upon administrative efficiency.

Improved telecommunications do, undoubtedly, accelerate the trend toward globalization of businesses, financial markets, and trade. Stock dealing can take place globally round the clock; goods can be ordered and shipped more efficiently and accounts can be settled rapidly around the world.

Some risks of varying severity were identified. One is the risk that drivers might have their attention diverted by conversing on the mobile telephone, and their control of the vehicle impaired by using one hand to hold the phone. This risk can be averted, if the law-makers wish to do so, by suitable regulations, supported by technology.

Some people are worried about the risk that the state, should it become malevolent, could use the possibilities offered by electronic exchanges for monitoring telephone conversations and other communications. Others do not fear the malevolent state and see these possibilities as useful in the fight against crime and in offering the consumer more detailed billing.

Another worry is related to the trend of removing telecommunications from the public domain and transferring it to the business domain. This raises the fear that the principle of the so-called universal service, which obliges the PTTs to provide all those who want it a telephone service at an equitable and affordable price, might be eroded or abandoned. A purely commercial service provider has no incentive to provide such universal service and might penalize customers in remote areas, or small consumers. The principle can be safeguarded only by the state acting either as owner or as regulator.

Some voices drew attention to dangers to health from extensive work on visual display units (VDUs). It is hard to tell how much of this conjectured danger is real. No doubt increased use of inter-computer communications will increase the amount of time people spend in front of their screens and this would add to the health risks, if they truly exist. Technologists suggest that a properly designed VDU does not present a danger to health.

Step 5: policy analysis

Policy analysis is the centre-piece of this technology assessment. The future of the telecommunications industry is dominated by questions of policy. In particular, it is government regulation of the industry that is crucial to its future, but the individual operator's policies are also of considerable importance. In the case of this particular client, being a virtual monopoly operator in the public domain, its influence on public policy is very considerable.

In the past, the PTTs had a complete monopoly, not only over the network and its equipment, but also over customer premises equipment. This situation is clearly untenable. With a huge variety of CPE on the market, the operator could not sensibly claim a monopoly over its sale. In the past, all equipment to be used in the network had to be approved by the operator in order to ensure the integrity of the network. This approval system was open to protectionist abuse. Sooner or later the European Economic Communities (now the European Union) will have to arrange for common European type approvals and for some form of standardization of equipment and operations.

The question of PTT monopoly has become acute because an anti-monopolistic stance is one of the new tenets of political faith. It used to be thought that a telecommunications network, with its miles of buried wires (or wires suspended on masts) and its central exchanges constituted a so-called natural monopoly, meaning that it would be wasteful to duplicate such a network. But with the advent of alternative forms of transmission, and with the idea (pioneered in the US) that a service operator need not own a full network but could be given the right to buy capacity from another owner of a network, and with the ascendancy of a new economic liberalism, the idea of a natural monopoly has been largely abandoned. The new ideas demand competition and believe that competition brings down prices, thus increasing demand and causing economic growth. The report therefore deals extensively with a description of liberalization, and so-called de-regulation, which in reality amounts to re-regulation in the US and other countries, and speculates about the form re-regulation might take in the client's country.

Though the principle of universal service is left more or less intact, competition among telephone operators has been introduced in some countries, with the US and the UK in the forefront of such developments. We have seen that technically the distinction between various forms of information is disappearing: it is all becoming streams of digits. Because of this technical trend and the political desire to remove monopoly, increase competition, and hand over all commercial operations to private enterprise, there is an increasing trend toward lumping together all forms of information being supplied to, or transmitted from, households and businesses. It is possible that television cable operators will be able to supply telephone services, and that telephone operators may supply television programmes.

Finally, policy options for the client are set out. The first option is to do nothing, though this might result in unfavourable treatment by the government. The next option is to start an own policy of liberalization. The first step would be to introduce universal sockets that allow all types of customer premises equipment to be connected to the network. The next step would be to simplify type approval, possibly in cooperation with other firms and the authorities in the EU. Tariff adjustments are also possible, in an attempt to make prices of services reflect costs more closely, though full reflection of costs is difficult to achieve as overheads are not readily attributable to individual services. The client does not believe that the principle of universal service will be abandoned, hence the policy response to such an eventuality was not analysed. It is regarded as very unlikely that the client's country will follow the path of extreme liberalism and introduce a great deal of competition, with its very mixed blessings, in the provision of basic telecommunications services.

On the other hand, higher services (known at the time as value added services) may well be thrown open to competition. Mobile telephony is already operated by two rival companies and the possibility that the client will be forced into selling capacity to rival service providers cannot be disregarded. Although the client will try to defend the present network monopoly, a fall-back position of negotiating a price for allowing other operators to use the network needs to be prepared. The option of not preparing such a fall-back position is discussed, as well as the option of giving up the monopoly voluntarily.

The policy options of service provision are analysed. With so many new services becoming possible, the options are either to attempt to provide them all, or provide only those that are closest to the present business and leave others to specialist providers of so-called value added services. This could leave the provision of basic services and, possibly, a few selected services, entirely in the hands of the client, but it could also open the way toward competition in a range of services. There is a danger in not providing value added services: the client might be left with only POT and lose the opportunity of developing more lucrative forms of services.

Although not strictly part of the brief, equipment purchasing decisions are briefly discussed. Particularly the costs and benefits of the use of optical cable, of microwave transmitters and of satellite capacity are mentioned. There is a discussion of the future of the so-called integrated broadband communications (IBC) network. It is possible, and the EC is championing this cause, that a universal network will be constructed, enabling the provision of interactive services, video services, multi-media services, electronic mail, and high definition television on the same network as ordinary tele-phony. There are several alternatives to this integrated broadband network. For one thing, compression technologies reduce the need for bandwidth. Second, there are ways of providing bandwidth on demand, reducing the total need for network capacity. The policy options are either to join one or another of these options or to play a wait-and-see game and develop the network as required by real demand. This option makes it necessary to be very flexible and fast in investment decisions and to reduce lead times.[2]

---

2 For actual technology assessments dating from that period see e.g. Ungerer and Costello 1988; CEC 1989; Braun 1990; US Congress, OTA 1990. For more up-to-date analysis see e.g. Mansell 1994; Cas and Pisjak 1996.

3

STRATEGIC MANAGEMENT
OF TECHNOLOGY

Introduction

The present text deals with technology assessment as an information input into the management of technology. We have seen that one of the most frequent tasks of industrial technology assessors is to prepare and analyse the technology component of strategic plans. As technology assessment is a lengthy and wide-ranging process of information gathering and analysis, its main role must be in the long fundamental, that is strategic, management of technology. To make sense, we must therefore clarify what we mean by strategic management and we must justify our earlier assertion that the management of technology is a specialist task that should not, in large firms, simply be subsumed under general management.

It is often said that top management pays insufficient attention to technology in its strategic thinking. This may be the case, but as I think that technology is at the very core of many business enterprises – in almost all manufacturing industry and in many service industries – I shall assume that top management gives technology all the attention that it deserves. This will be the underlying assumption in our further discussions of strategic management. And, if this is the case, the role of technology must be viewed strategically, as a vital component of the basic competitive stance the firm wishes to take. We must therefore view technology in as broad and long-range a fashion as is possible. In other words, we should apply the methods of technology assessment.

The twin enemies of successful strategic thinking are myopia and tunnel vision. Technology assessment is an instrument to remedy these defects: it widens and extends the field of vision – at least in matters technological.

Strategic management is not a completely determined and non-controversial subject. We shall not, however, quote the full range of definitions and attitudes to strategic management found in the literature. Instead, we shall adopt a particular stance, without thereby claiming that our point of view is the only legitimate or useful one. The student who wishes to consult a critical review of writings on strategic management is referred to Whittington (1993) and to Mintzberg (1994). The former clearly describes the different schools of thought on strategic planning and critically discusses their ranges of applicability.

There is a great deal of discussion in the literature (e.g. Whittington 1993) about the ultimate goals of strategic planning. In my view, all strategic planning must serve the ultimate purpose of survival (unless the firm is suicidal or wishes to be taken over), but many different fundamental long-term competitive stances and many intermediate strategic aims are imaginable.

The firm needs strategic management for long-term survival and prosperity. Technology is vital to the life of the firm and is one of the most important tools available for taking up a certain strategic stance. Thus technology needs specialist strategic management. Strategic management of technology requires an information input in the form of technology assessment.

Strategic management

Strategic planning, distilled to its essence, is an attempt to steer a large organization toward a desired goal, instead of allowing it to be buffeted hither and thither by external forces. The steering can have only limited success, as external forces can neither be eliminated nor entirely predicted. Yet some success is attainable and is worth the effort.

Strategic management is shorthand for two linked activities: the formation of a strategic plan, and the implementation of this plan. The implementation is more akin to normal management activities, because it requires actions. The formation of the plan, the activity of strategic planning, requires no action; it is an exercise in the gathering of information and in policy formation. A strategic plan lives on paper; its implementation requires real investments, real purchases, real activities. As we are dealing with technology assessment, and this activity is part of strategic planning, we shall concentrate solely on planning rather than on implementation.

Whittington (1993, 10–41) describes four fundamentally different attitudes to strategic planning: the classical approach; the evolutionary perspective; processual approaches; and the systemic perspective. The last of these approaches seems to be particularly appropriate for the present time. It retains some faith in the possibility of organizations acting rationally in planning their future, but it asserts that the rationality and 'objectivity' of individuals and organizations is conditioned by the environment in which they are embedded. What appears economically rational depends upon social and political conditions, upon the Zeitgeist, upon a web of personal relationships, upon apparently self-evident tenets of faith and tacit assumptions. This approach fits in well with our approach to technology assessment, as the technology assessor also has only limited potential to rise above his/her time and see the world with 'absolute' objectivity. The rationality available to us may be bounded by our circumstances, but we must not forget or neglect the distinctions between logical conclusion and non-sequitur, between judgement and prejudice, and between pure and enlightened self-interest. Without these, all rationality – indeed all civilization – goes out of the window and every whim becomes as good as a rational argument. We must beware of following fashions blindly, and must maintain a rationally critical attitude – even if we cannot entirely escape the conventional wisdom of our time.

Multi-national corporations vary in their behaviour according to their main national base, as they are dominated by the attitudes of their home-based top management. 'Thus national approaches to strategy can be heavily distorted by what is locally regarded as culturally legitimate' (Whittington 1993, 30–31). It seems highly probable that the attitude of top management to technology as a strategic asset depends upon these cultural differences; Anglo-Saxon managers probably care less about technology than their German or Japanese counterparts.

> Thus, attention is drawn to the way in which product-market success increasingly depends upon the focused acquisition of technological knowledge through interfirm collaboration. The building of competences rather than the financial management of portfolios is seen as the critical strategic task. Much of the thinking behind this view is based upon the success of Japanese organizations in developing interrelated technological competences.
>
> (Scarbrough and Corbett 1992, 149)

Technology is selected within an organization in ways that suit its organizational concepts and its dominant views and goals (organizational culture). On the other hand, the technology itself influences and limits choices of organizational features. Organization and technology are interdependent and the technology assessor, for all his/her subservience to top management, does influence the thinking of top managers. Neither technology nor organization are truly independent variables.[1]

The strategic plan, once adopted, represents a set of decisions to be implemented. 'Planning is a formalized procedure to produce an articulated result, in the form of an integrated system of decisions' (Mintzberg 1994, 12). The strategic planning process involves both internal and external appraisal. It is essentially based on the so-called SWOT (strengths, weaknesses, opportunities, threats) model in that it analyses the internal strengths and weaknesses of the firm, as well as the opportunities and threats in the outside world. The planning process makes use of concepts such as gap analysis, i.e. it looks for gaps in the technological capabilities, in the product range, and in the firm's markets. The planning process may be highly formalized or fairly informal, and it may use a variety of techniques. Once a plan is adopted and has become a set of decisions, it sets out a desired range of products and desired markets for the firm. In other words, the strategic plan defines what business the firm wishes to be in.

The strategic plan represents a set of goals to be achieved. It describes the state the firm wishes to be in at some future time, and sets out the steps necessary to achieve this future state. The total plan consists of several sub-plans for different areas of the firm's activities. It is of the utmost importance that the sub-plans should be integrated to mesh together without contradictions to give a self-consistent total. The technology plan must provide the right technology to produce the right products at the right price. The entire organizational structure of the firm may need to be adjusted to new aims, new acquisitions, new products and new markets.

The important feature of a strategic plan is that it gives the firm a detailed set of aims, and thus a means of controlling its achievements: 'much so-called strategic planning activity reduces to not

---

1 For a detailed discussion of the relationship between organization and technology see e.g. H. Scarbrough and J. M. Corbett (1992). See also *Technology Analysis and Strategic Management* Vol. 7, No. 3 (1995), guest editors K. Dickson and A. Genus.

much more than the quantification of goals as a means of control' (Mintzberg 1994, 54). The reason why we speak of strategic management, rather than just strategic planning, is that the process of planning and the process of implementation of the plan must be closely related. It would be possible to leave the task of implementation of strategic plans to line managers, except that these tend to be overloaded with short-term problems and ad hoc decisions, which might leave the longer-term plans in danger of ending up on the back-burner. If today's demands are pressing, tomorrow's demands get shelved. It is therefore sensible to give the task of implementing the long-term plans to those who are not too burdened with today's problems. It is hardly necessary to stress that constant consultation and coordination between line management and strategic management is vital, as otherwise their inherent conflict of interests could easily get out of hand.

The process of strategic planning requires a great deal of data gathering and of analysis. Planners have to gather facts about consumer attitudes and buying habits, about changes in the behaviour of competing firms, about changes in the regulatory regime and changes in world trade patterns. Ideally, the planners should know about likely costs of capital, exchange rates, and investment support available in different countries, prices of raw materials and energy, labour costs, etc. Last, but not least, they have to gather information about forthcoming technologies and products. What new production technologies are forthcoming, what new products can be foreseen, what are the likely trends in consumer behaviour in different markets?

Much planning is done on an ad hoc basis; special studies undertaken for a variety of reasons, such as perceived major changes in technology, in the environment of the firm, or in its management. Some writers claim that these one-off exercises are more important to strategy formation than the routine annual plan.

Strategic planning clearly is concerned with the future. It attempts to foresee the future, but also tries to equip the firm with the means to weather unforeseen changes. This can be achieved either by rapid response, i.e. great flexibility, or by covering a large range of products, markets and technologies, so that there is a chance that even if some suffer setbacks, others will thrive. Small firms, by their nature able to be flexible but unable to cover a broad field of activity, are forced to rely on flexibility; large firms, less able to be flexible but more able to cover wide fields of activity, tend to rely on coverage.

It is hardly conceivable that a strategic plan should not have a major technological component. Yet if strategic management of technology is to be effective, the most senior strategic technology manager has to be located at a very high managerial level, so as to be directly involved in decisions about the firm's strategic plan and in the supervision of the execution of the technical aspects of the plan.

The lower levels of strategic technology management have three functions to fulfil: information gathering and analysis; preparation of planning proposals; and implementation of plans. The first function is identical with major aspects of technology assessment. The second function is, in the political arena, also part of technology assessment and there appears to be no good reason for not combining these two functions into a single technology assessment task. Thus the gathering and analysis of information would be combined with the preparation and analysis of planning proposals. This is consistent with the TA methodology – scope, technology, impacts, policy (STIP).

The implementation of the technical aspects of a strategic plan, adopted by senior management, can be carried out by a different group of technology managers. It goes without saying that all groups need to cooperate and to coordinate their activities. It may also be wise to move people between tasks, thus broadening their experience. Perhaps it is salutary for a technology assessor to put his/her plan into practice. Some feedback of information between technology assessment and the implementation of plans is necessary in any case.

Top management in commercial enterprises plays the role of elected politicians in the public arena. Technology assessors in commercial enterprises play much the same role as in the public arena: they supply information and analyse policy options. Middle management plays the role of civil servants in implementing policies decided on by top management or politicians, as the case may be.

Many organizational forms are possible and may work equally well; what is important is to ensure close proximity between TA analysts and strategic managers. There also needs to be close contact between top management and strategic technology management, to ensure that the technology plans mesh in properly with the general goals of the firm.

Much information resides with people and is not available in any formal processes. No matter how many computer information networks exist and how well they may be organized, the informal exchange of views and opinions and snippets of information is, and will probably always remain, indispensable. For what people say to each other in informal exchanges is very different from what they commit to paper or computer memory. Informal exchanges both between collaborators and rivals form an important aspect of the total information on which economic and political players act. The only ways in which the exchange of soft information can be improved is by allowing informal networks to thrive and by encouraging wide-ranging consultation as part of the information gathering process.

Before discussing strategic planning and management for technology any further, we must try to understand the role of technology in commercial firms in a little more detail.

The role of technology in commercial firms

Technology, as defined in this text, plays several roles, dependent on the line of business of the firm; all of them vital and many of them of strategic significance.

First and foremost we need to mention *production and process technology* in manufacturing firms. The way the firm produces its products determines the quality of the product, the production cost, the required skills, maintenance needs, cost of materials, and production capacity. It further plays an important role in the safety standards and in the environmental hazards both inside and outside the manufacturing premises. Finally, the availability of production technologies, including the skills and the organization required to operate them, determines the range of products the firm is able to manufacture. It also determines the flexibility of production, meaning rapid changes in the product mix and fast product innovation. All in all a formidable list, considering that the range of products, their quality and their cost are the major determinants of a manufacturing firm's competitive position.

Modern manufacturing technology has become a highly sophisticated mix of computer control, automated machines, a web of closely linked suppliers, a multi-skilled workforce, built-in quality control, and a well-honed organization.[2]

In some branches of industry production technology is the dominant factor in competitiveness. In the automobile industry, for example, in which there is very considerable convergence of the product and very

2 For details the reader is referred to the literature, for example Bessant (1991).

fierce, albeit oligopolistic, international competition, the competitive position of the firm is largely dominated by production technology and, hence, quality, cost, and flexibility. For a fascinating account of the international automobile industry and its 'lean production' methods see Womack *et al.* (1990). It is the production methods pioneered by Toyota and other Japanese manufacturers that has given the present techno-economic paradigm the name 'post-Fordist'. In the semiconductor industry, on the other hand, we have seen that advances in the product are intimately linked to advances in manufacturing and process technology. What you can produce depends not only on what you can design, but also, crucially, on how you produce it. The cost of production of silicon chips is also dominated by production technology, particularly because it determines the production yield – the ratio of acceptable products to rejects.

Infrastructural and ancillary technology is also important to manufacturing firms. By this we mean all the ancillary technologies such as heating and ventilation, air-conditioning, internal transport, internal and external communications, office technologies, warehousing, power supplies, water supplies, sewerage, and so forth. A firm may be good at producing its products, but if its office technologies, its communications, its production control systems or other ancillary technologies are outdated and inefficient, the firm will suffer. Even sales and service departments now use sophisticated computer and communications technologies, and this can give the firm a real competitive advantage.

It goes without saying that the *product* is crucial to the life of a manufacturing firm. The choice, design, technical capabilities and features of products, as well as their quality and price are essential to the competitive position of the firm. Even when the product is, say, a garment, its quality and price are dictated by design and by technological choices: the choice of materials, the choice of production technologies, the choice of finishes. If the product is a technological device, say a refrigerator or a compact disc player, the technical qualities of the product are of the essence. Does the refrigerator use refrigerant gases that contribute to the greenhouse effect? How good is its insulation? How efficient is it? Does it defrost automatically? How noisy is it? How long will it last? How well organized are its different temperature compartments? These are questions a well-informed consumer asks, in addition to considering appearance, ease of cleaning, convenience of food and drink storage, and, last but not least, price. For a compact disc player, apart from appearance and price, the concerns of the buyer are the quality of sound reproduction, the ease of use and of programming, the range of useful programming possibilities, longevity and ready availability of servicing and repair facilities. Even this brief discussion of just two examples should amply demonstrate that the technological qualities of products are vital to the success or failure of the product. The only thing that limits the dominance of technical features is the fact that consumers find it very difficult to judge them properly, and hence reputation and image, and thus salesmanship, play an important role in determining the fate of products in the market place.

We have enumerated some features that are in the consumer's mind when buying certain products, but the features the consumer considers are subject to change and to fashion. Technological innovations are introduced in order to stimulate new demands by consumers and, if the innovation is successful, firms other than the innovating firm need to imitate it rapidly if they are not to lose market shares. Though the introduction of innovations – technically new or substantially changed products – is a gamble, not to imitate successful innovations is a death warrant. At any given time, each manufacturer's products within a certain price range must equal or better the technological specifications and performance of other manufacturers' products.

Many texts proclaim that technological innovations are introduced in response to consumer demands – so-called market pull. We do not deny that some ideas for innovations are a result of consumer research or may be stimulated by retailers or sales-people; yet we maintain that most innovations are driven by technological possibilities, linked to expectations of consumer acceptance – so-called technology push. In firms supplying parts to assemblers and in firms producing production machinery, the buyer has, of course, a great deal of influence upon innovation and a situation akin to market pull may well arise. But whether the innovation is triggered by market pull or technology push, the result, in terms of the need to keep products up to current technological standards, is the same. Products are subject to fashion, in specification and performance as well as in appearance, and manufacturers will ignore the demands of fashion at their peril.

Technologists and scientists like nothing better than to produce technological innovations. In many ways, it is the very essence of their existence, for if technology remained static, fewer technologists would be needed and much of the challenge would go out of their professional existence. This is a major reason why the management of technology cannot be left to technologists alone. They would produce a constant stream of innovations, always hopeful that the market would accept them, and the firm's profits would flow down this stream. The negative cash flow of the early stages of an innovation would never be given a chance to turn positive in the consolidated stages of the new product. Technologists might provide the dynamic drive a firm needs, but they have to be closely watched by guardians of financial sanity. The creative tension between innovation and consolidation, between risk and caution, between technologist and accountant, is vital to a firm's health.

For large firms with their own R&D departments, we may speak of R&D strategy as an additional component of strategic technology management. The capabilities of the R&D department often play an important role in the overall technological capability of the firm, and the R&D department is the source of much of the innovative activity of the firm. This is where ideas and developments for new products come from; and, quite frequently, it is the firm's own R&D that makes important contributions to new manufacturing technologies and processes. R&D departments often maintain a scientific-technical knowledge base which enables the firm to absorb new products and new processes much more readily than it could without this base. In some ways, R&D departments are gate-keepers watching out for scientific developments, and are a storehouse of knowledge that may be of great practical use some day. R&D can serve all these functions on two conditions: that there are close links between the R&D department and the rest of the firm; and that R&D is not conceived and constrained too narrowly. However, R&D strategy and R&D management are topics of such great significance that they have their own research agenda and literature. To us, it is a marginal issue and a few remarks on it must suffice.

There is yet another source of demand for technological updating of products – the pressure of regulations. As new hazards caused by the use of certain products come to light, as new knowledge about dangers to health or environment is discovered, so regulatory demands on products change. Compliance with such regulations, that may be different in different markets, is mandatory. Ignoring regulations is not an option for manufacturers, unless they wish to withdraw a particular product from a particular market. Take the motor car as an example. In recent years regulations about exhaust emissions have been tightened considerably, making it impossible in many markets to sell petrol-engined cars without catalytic converters. Regulations about passive safety have also been tightened, so that cars must be able to protect their dummy occupants in prescribed crash tests and must be equipped with seat-belts for all passengers. Public opinion has shifted a great deal in the matter of car safety. While in the past most people ignored safety aspects of vehicles in their purchasing decisions, many now rank such considerations before all others. This has compelled manufacturers not only to comply with regulations, but to be seen to better them. This is an instance when the market, stimulated by a few pioneering manufacturers of quality cars, has signalled innovative demands to laggard manufacturers. It is also a case when regulation, plus information about vehicle safety made available by the authorities, aided and abetted by pressure groups, have greatly contributed to a reduction in the number of victims of road traffic.

A greater consciousness of environmental issues in some sections of the public has prompted some manufacturers to try and establish a 'green' market niche for certain products, such as household cleaning products. Unfortunately, these products are sold at premium prices without any guarantee that they are environmentally more benign than their rivals. There is no substitute for environmental regulation. The least the regulator can do is to compel manufacturers to label their products accurately and informatively, thus enabling the consumer to make a rational choice. For labelling to be effective, however, objective scientific knowledge about the environmental effects of the product needs to be made available.[3]

Technology in service organizations might seem, at first sight, to be less central to the life of the organization than technology in manufacturing firms. The degree of importance of technology in service organizations is variable. Many organizations, such as purveyors of pizzas, providers of cleaning and catering services, and many others get by with relatively simple technology. For the majority of service providers, however, technology is extremely important, and for some it is quite crucial. Some organizations provide services based almost entirely on information handling. Prime examples are banking and financial services, insurance, travel and holiday agencies, a miscellany of consultancies, news gathering services, newspapers. These companies rely very heavily on information technology – computers and telecommunications – and their competitiveness depends to a

---

3 For a case study of eco-labelling and energy labelling in the European Union, and its problems, see Potter and Hinnells (1994).

considerable extent on their skill in deploying and handling these technologies. Even retailers now depend heavily upon technology. Quite apart from refrigeration and such like, their check-outs, their stock-keeping and re-ordering, and their controls over profitability of product lines, all depend on information technology.

Many important service providers are entirely dependent upon a wider range of technologies. Airlines, railways, bus companies, health services, all these depend upon technology as much as do manufacturing firms. It is vital for an airline to order suitable aircraft at the right time; it is equally vital to adhere to rigid and efficient maintenance schedules, to deploy spare parts at the right time in the right place, to maintain excellent reservation systems, to use technology effectively for training, and for flight and crew scheduling. Transport undertakings may indeed be regarded as purveyors of technology; their function is to retail transport technologies to the public. The health services also rely heavily on information technology for keeping patient records and other administrative functions, and they rely, of course, upon the whole gamut of medical technologies, whether in the operating theatre, in diagnostics, or in therapeutic procedures.

As is well known, Michael Porter (1985) divides the activities of firms into value chains in order to pay special attention to the enhancement of value creation. As to the analysis of technology as a strategic tool, he recommends seven steps (Porter 1985, 98–200):

1 Identify all the technologies in the value chain. He stresses that not only are the main technologies important, but that ancillary technologies may also play an important role. The analyst should know the role of different technologies and their contribution to the value created in the firm.

2 Identify potentially relevant technologies. This means looking at rival firms and at emerging new technologies and considering their relevance to the firm.

3 Consider the likely development paths of all relevant key technologies. As has been stressed several times, the future development of technologies must be considered in order not to miss the train, as it were.

4 Consider which technologies are most important determinants of competitive advantage. This involves considering which technologies create sustainable competitive advantage; which shift cost or differentiation in favour of the firm; which might

give advantages of technology leadership; and which improve the overall structure of the industry.

5 Assess the firm's relative technological capabilities and the means and cost of improving them.

6 Select a technology strategy, encompassing all important technologies, that reinforces the firm's overall competitive strategy. This includes ranking of R&D projects according to their significance for competitive advantage; choices about areas in which technological leadership is sought; policies toward licensing; and means of obtaining necessary technologies from outside.

7 The seventh point applies to large corporations only and asks for business unit technology strategies to be reinforced at corporate level.

Strategic management of technology

Drejer (1996) points out four different tasks of technology managers, referred to in the paper as historically developed schools of management of technology (MOT). The schools, or tasks, are: R&D management; innovation management; technology planning; strategic MOT. One important task is left unmentioned in this list: the selection, fine tuning and maintenance of efficient production machinery and processes. This is usually the domain of the production engineer, but now includes functions such as management of computer and telecommunications services.

R&D management has always been in the hands of scientists and technologists. It includes the difficult and crucial task of selection of research projects which, to be done properly, ought to be closely related to the strategic plan of the firm. Though some very large firms can afford the luxury of telling their R&D laboratory that they should carry out research in all spheres that might be useful to the firm's range of activities, less fortunate firms have to be more specific in their research targets. The age-old questions of how fundamental, or how applied, industrial research should be will presumably always have to be resolved on a case-to-case basis. Though R&D laboratories sometimes become involved in trouble-shooting and are asked to look into causes of manufacturing failures, essentially their task is to produce ideas for innovations and develop selected projects to a stage where the manufacturing branch of the firm can take them under its wings. Details of organization vary, the important thing is to keep close contact between R&D and manufacturing and, of equal importance, between R&D and strategic planning.

Innovation management consists of many related tasks. The innovation managers must make sure that the process of selection of innovation projects is efficient, that the development process of a selected innovation proceeds as required, and that early preparations for the introduction and/or manufacture of the new product or process are made. Finally, the innovation managers must nurse the new product through the turbulent and chaotic early stages of production and feed the marketing organization with all necessary information. Marketing should reciprocate by giving feedback and suggestions from the market. Innovation managers may have to coordinate many feedback loops in the process, such as requesting further R&D if necessary.

Technology planning comes close to technology strategy as it means planning the range of production technologies and the range of products the firm should aim for in the longer term. This involves scanning the horizon for new technologies and observing rivals very closely. It is in the selection of innovations and in the selection of technologies that technology assessment is of particular importance. Both these aspects are part of strategy, and both should be carried out with thought not only for the immediate likely financial returns, but also for the impacts upon the workforce in terms of employment, skills, hazards and prospects. Thought ought to be given to environmental impacts and social utility. By social utility I mean the degree to which a technology serves environmental improvement, prevents environmental degradation, serves social purposes such as reducing congestion in cities, improving public health, and so forth. In the selection of innovations care ought to be exercised not to innovate faster than necessary. It is necessary to make sure that the innovation offers real improvement, not just novelty. Similarly, in manufacturing innovation the selection process should not aim necessarily at reducing the labour input, but should consider carefully where automation is useful and where the flexibility, skills and creativity of people are major assets. Considering all these aspects in the selection of innovations and in the selection of production technologies means employing technology assessment. It is technology assessment that takes the long and broad view, instead of the narrow myopic view.

The road from R&D management as the only task for technology managers, to the assertion that technology management ought to be carried out at strategic level, reflects the attempt of technology managers to gain access to the higher echelons of management (Green *et al.* 1996). It also reflects the increasing recognition of the importance of technology in the life and fate of commercial enterprises in the modern world.

We have said that the end product of strategic decisions is a selection of products and markets for the firm, together with a timetable and a set of steps to be taken for reaching this position. We have also said that the products and their potential success in the market place depend heavily upon production technology deployed and upon technical features of the products. In service organizations, on the other hand, success of 'products', or service packages, also depends to a variable, but not inconsiderable, degree upon the deployment of suitable technologies. Thus technology inevitably must play a role in the drawing up of a strategic plan – in setting strategic targets – and the acquisition of suitable technologies must form part of the steps to be taken in implementing the plan.

It is just as impossible to implement a strategic plan without taking the right steps toward the acquisition and operation of the right technologies, as it is to implement a plan without proper financial provision. Indeed it is impossible to plan products without paying attention to their technological features and to the methods of manufacturing and assembling them. Technology is as important an ingredient of strategic planning as are financial and market planning. The self-consistency of a strategic plan can only be assured if all aspects – personnel, finance, acquisition of buildings, choice of products, development of markets, and technology – mesh together smoothly and without contradiction. There are many feedback loops within the planning process; particularly, the need for finance depends critically upon technology and product planning, but also on market expansion plans and on planned volumes of production.

Strategic management of technology means the planning of the development or acquisition of production technologies and of products with the right technological features, to fit in with the general strategic plan of the firm. Strategic technology management also means implementing the longer term technological plans. It does not mean the day-to-day running of production or maintenance, which falls onto the shoulders of technology line management.

When we speak of a firm, we do not necessarily mean a large conglomerate, but what amounts to a separate business, even if it is a strategically almost autonomous sub-unit of a larger firm. We note in passing that the division of the firm into strategic units (or cost centres, or business units) usually occurs on the basis of two criteria: the product-markets the unit serves, and the technology it employs. Each business unit should have a homogeneous set of key factors for success and, at least in manufacturing industry, technology invariably is a key factor (Dussauge *et al.* 1992, 26–33).

Properly we should deal with the strategic management of production technology and of products as separate issues. Indeed, product planning is only in part an issue of technology; many aspects of product planning are the concerns of design, marketing, production capacity, competition, and coherence of product policy. To some writers on technology management, the technology of the greatest strategic significance is production technology. Bessant speaks of manufacturing strategy as an important component of business strategy, and regards production technology as a crucial aspect of manufacturing strategy.

> Manufacturing strategy is primarily concerned with how the products will be made, or the services delivered, and includes factors such as: choice of process or technique; make or buy decisions; setting quality standards and procedures; production and work organization; requirements for physical facilities (buildings and services); design of planning and control systems; investment plans and justification. One of the most significant tools in manufacturing strategy is technology, the combination of equipment, software and organization which facilitates manufacturing.
>
> (Bessant 1991, 15–16)

We are somewhat careless about keeping product technology and process technology separate, because they are intimately related. Related not only because the qualities of a product are largely determined by the way it is made, but also related in terms of their evolution in the process of technological innovation. Consider the evolutionary succession of technologies for a radical product innovation: we proceed from idea to innovation, from an innovative product to a mature one, maturity eventually giving way to obsolescence, leading to the decline of markets and possible demise of the product. Even during the obsolescence phase the product is succeeded by the following technological innovation. Abernathy and Utterback have shown that this cycle of product innovation is associated with a cycle of process innovation. They regard the early phase of a radical product innovation as the fluid phase, 'in which a great deal of change is happening at once and in which outcomes are highly uncertain in terms of product, process, competitive leadership, and the structure and management of firms' (Utterback 1994, 92). In other words, this is a turbulent phase in which management has to be, above all, highly flexible. Although the innovation is likely to be part of a strategic plan, its exact progress in the early stages of production is unpredictable and the firm has to dive through the turbulence as best it can. The product designs are still diverse and the production technology is flexible and inefficient. The next phase is the transitional phase, in which the product changes slow down and a dominant design begins to emerge; production technology is being developed rapidly to accommodate the rising demand for the product. In the mature phase, the product becomes more standardized and the production technology becomes highly efficient (Utterback 1994, 80–99). Thus we see that product and process innovation are linked and, in those phases of production where orderly strategic planning is possible, i.e. in phases where technological change is incremental rather than radical, product and process strategies must be coupled.

Sometimes a radical innovation consists of a major improvement to a mature and established product. The process has been termed re-innovation (Rothwell and Gardiner 1989) and aero-engines provide a good case in point. Utterback's basic argument should hold even in this case.

The strategic technology manager plans and implements the technological features of future products and plans and implements the production technologies for the manufacture of these products. Some, probably most, of the technological requirements will be satisfied by outside purchases, others will be developed within the firm. Sometimes technological requirements are prime considerations in take-overs. If you need a technology that another firm possesses, one of the ways of obtaining it is to purchase the firm. Occasionally firms are taken over by others for purely commercial, non-technological reasons. Yet the result may be a fortuitous major gain in technology (S. Macdonald documented such a case in a large British company, to be published in the *International Journal of Management of Technology*). Short of taking over a firm, there are possibilities of forming alliances with other firms for the development of technologies (Braun 1995, 64–76).

The introduction of new machinery, new controls, new organization, or new layout into a manufacturing plant is a complex innovative process in its own right, even if all the new ingredients introduced into the firm have been tried elsewhere. The process has been termed manufacturing innovation (Braun 1981). For the equipment manufacturers it is part and parcel of the process of diffusion of their innovative technology, but for the manufacturer using the methods for the first time, it can be a major innovative effort. It is undertaken if in the process of constant self-critical examination the established manufacturing procedures are found wanting, or if a new product is to be manufactured. A search for a solution is then instigated, and when a particular solution is decided on, the process of implementation begins. The self-critical and search phases can and should be part of the technology assessment procedures. Manufacturing innovation is a four-phase process: the need for improvement is discovered; the actual weak links are analysed; solutions are sought and decided upon; solutions are implemented. It is a complex process, and for it to succeed, management must see to it that a constellation of propitious circumstances is created.

If the introduction of new production machinery – we are now speaking of individual items of equipment rather than of whole new production facilities – is regarded as the diffusion stage of an innovation created by the equipment manufacturer, then it is customary to view the early purchasers of the new equipment as pioneers, later purchasers as followers, and very late ones as laggards. Obviously the pioneers take certain risks with untried equipment and may be called upon to help the manufacturer put the finishing touches to the development. On the other hand, they gain the benefits of almost tailor-made equipment and of becoming early users of improved manufacturing methods. The laggards, on the other hand, run no risk, but will probably pay cost penalties for using obsolescent equipment longer than others. There is no general certainty about which is the best position to be in – it is all a matter of specific circumstances, such as how old the old equipment is – but generally the choice between being a leader or a laggard is a strategic choice.

Technology and strategic positioning

A strategic plan is formed on the basis of decisions about strategic positioning of the firm. A firm can take many different strategic stances, though because of its history, its existing structure, and its markets, not all possible options are actually available to it. The choice of strategic position mostly dictates technological choices or

71

vice versa; technological choices can dominate the choice of strategic position.

The two main groups of strategies are cost leadership and product differentiation.

Cost leadership

If a firm decides to try and achieve a competitive advantage through low cost, meaning that it must attempt to lower its overall costs as much as possible, technology invariably is a major weapon in achieving this goal.

Costs are, of course, determined by a great many factors, not all technological. The cost of capital, the general efficiency of the organization, the state of the market, are all important factors. Technology affects costs in three ways: the cost of depreciation of machinery and equipment; the productivity of the production process; the design of the product. The first of these is not really under the individual control of the firm. If it happens to employ machinery that is rapidly becoming obsolescent, it will hardly be able to keep such machinery and still be a cost leader.

The efficiency of production itself is very much under the control of the firm. By employing the best available machinery and the best possible organization of production, by using skilled workers and giving them incentives and opportunities to suggest improvements and see to quality throughout the manufacturing process, by maintaining all the equipment in perfect shape, by having a time-saving layout of the production facilities, by employing reliable suppliers of components who deliver perfect quality just in time, thus obviating the need for quality control of bought components and the need for keeping large stocks, by designing the product for ease of manufacture and using all the other appropriate ingredients of modern production management, the firm can increase productivity to the highest possible level. This brief paragraph tries to sum up what has been written in dozens of books and articles and has been developed over a good many years (see e.g. Womack et al. 1990; Bessant 1991; Rhodes and Wield 1996). It needs to be stressed that even the best production methods used in one location cannot necessarily be completely transferred to another. The cost structure of supplies and labour varies from place to place, the infrastructure varies, road congestion may, for example, make just in time delivery much less effective than it is in places with empty roads. It is one of the tasks of technology management to bring the production facilities up to the highest accessible standards and keep them there, remembering that the highest standards are a moving target.

Although methods of production have become much more flexible in the sense that it is now possible to achieve the highest productivity with shorter runs and with some variety in the product mix, it is still true that only a large volume of production allows the strategy of cost leadership to be effective. Economies of scale are still important, though they may not be quite as significant as in the past. There are two reasons for this. First, the fixed overheads have to be spread over a large volume of production. These include the production facilities themselves, the design of the product, R&D costs (if any), management and marketing costs, and so forth. Second, there is the well-known phenomenon of the learning curve. The efficiency of production of a new product increases for some considerable time with the volume of production. In other words, in the early stages of production the process has not been honed to a fine art. Some aspects of product or process may need alterations, some procedures may need to be changed, some software may malfunction. Much knowledge learned is in fact tacit knowledge – it resides in the hands, eyes and brains of the operatives and cannot be formalized.

One of the ironies of modern industrial life is that innovation is the killer of learning. We are compulsive innovators; and yet each innovation brings us back to the beginning of a learning curve (Dussauge *et al.* 1992, 46).

> There is much learning associated both with making a reorganization function and making the best use of new machinery. If both become obsolete rapidly, if indeed they are obsolescent at the time of introduction, then the optimum learning and adjustment will never be achieved. The technology will never be used to its full potential, the optimum on the learning curve will never be reached.
>
> (Braun 1995, 193)

Economies of scale are not outdated yet.

To innovate is always expensive, but sometimes not to innovate may lose market shares, and that can be even more expensive. The only possible general advice is: innovate only if you are pretty sure that you will reap considerable benefits, otherwise go for steady improvement.

The second major cost factor that is under the control of the firm is product design. Products of equal quality and performance can be different in the effort they require for their manufacture, in other words in their ease of manufacture (often referred to as manufacturability). Product and process have to be in tune with each other. This means reducing the number of component parts to a minimum, design for easy assembly, and so forth. There is an extra bonus to be had: design for easy manufacture often improves the quality and reliability of the product.

With some products no competition other than on costs and on marketing skills is possible. These are staple products, such as bulk chemicals, which are of the same quality and appearance the world over and where only good production facilities, good organization and good marketing can bring a modicum of success.

Product differentiation

Cost is, of course, nearly always important. But there is a range of strategies based on product characteristics that avoid some of the worst pressures for cutting costs. One way of achieving product differentiation is by creating a brand image, in the manner of washing powders. While on the subject of image, this is undoubtedly important. A recent example shows that a well-respected British firm felt it necessary to get rid of its 'metal bashing' image and acquire the image of a sophisticated modern engineering firm instead (Macdonald 1995). But we are not talking about pseudo-differentiation between brand names, we are talking about giving the product unique characteristics for which the customer is willing to pay a price.

A product differentiation strategy can be based on acquiring a reputation for excellence – reliability, luxury, highest technical specifications, superior design. This image must be based on real properties of the product, though advertising can help to make the image widely known and to reinforce it. Some car manufacturers have maintained this image for a long period; in other products it has sometimes proved difficult to maintain the differentiation in the face of technological developments. In optical instruments and cameras, for example, the high quality brand names of the past have been eliminated by computer designed lenses and products of virtually equal characteristics by a number of manufacturers. They now compete by giving their products more and more so-called features (whether the customer needs them or not).

Product differentiation can be maintained over long periods

either by using proprietary technology that is difficult to imitate, or by consistently excellent quality, or by being ahead in product innovation, or, sometimes, by acquiring a particular house-style.

If the product cannot be protected by patents or brand loyalty, then it must be protected by entry barriers.

> Firms which implement differentiation strategies must thus constantly strive to maintain entry barriers to the differentiated niche, create specific key factors for success and develop distinctive skills that ensure them a competitive advantage based on the mastery of these specific key factors for success.
>
> (Dussauge *et al.* 1992, 50)

The best chance for successful product differentiation lies in pioneering a highly desirable radically new product and giving it impregnable patent protection. If imitation is impossible, monopoly profits can be reaped throughout the life of the patent. This is a happy constellation that cannot be created at will. There is no magic wand that creates radically new products that are not only successful in the market, but are also inimitable. Indeed, trying to be a pioneer may bring great success, but it is also a very risky business. What if the product fails, either technically or in the market? What if protection can be breached? In many circumstances, particularly with products where patent protection is weak and/or licences are readily available, the fast imitator is better off than the originator. In any case, who comes first and who is second is often a matter of sheer luck. Sometimes it is a good policy to grant licences freely, as this is likely to increase the total market for the product and still leaves the holder of the original patent with a good market share and a good income from licence fees. This was the policy that caused the initial rapid spread of semiconductor technology in the 1950s, when licences and know-how for transistor manufacture were readily available from Bell Laboratories.

Some firms create their own production processes, which remain secret; some maintain leadership in pioneering products for long periods of time. Take, for example, Intel. They invented the microprocessor in 1971 and have maintained leadership in this technology ever since. Xerox grew into a very large firm on the basis of, initially exclusive, mastery of a single technology. Some firms produce a product of a highly specialized nature and maintain a market niche for it simply because the entry barrier is too high, and the niche too small, for other firms to try and enter. Some firms produce products of such complexity, say in aerospace technology, that the cost of entering is quite prohibitive and competition is restricted to a very small number of international players, with often exclusive products. For such firms their essential know-how resides in part in the R&D laboratories, but to a considerable extent also in their highly skilled staff (Dussauge *et al.* 1992, 53).

One of the many difficulties faced by firms attempting product differentiation strategies is that the true qualities of a product may not be readily perceptible to the public. When selling capital goods, the firm is usually faced with knowledgeable buyers and any technical differentiation will be perceived by the buyer and may, or may not, be worth a premium price. In this case the success or failure of differentiation depends on whether the buyer is willing to pay an extra cost for a particular technical property. In the case of consumer products, the general public is mostly unable to judge the true qualities of a product and retailers are often of little help in informing the public. This is where consumer organizations have an important role to play, though their efforts are often nullified by too rapid a change of models.

Product differentiation strategies can succeed in capital goods if the differentiated product offers genuine advantages to a section of the customer base, provided the differentiation can be maintained or constantly renewed. In consumer markets, differentiation can succeed if a brand image is established, probably with the aid of suitable advertising and help from retailers. In both cases, the retention of the advantages can be achieved by a combination of patent protection, secrecy, or continued technological leadership. Occasionally, a firm succeeds with a very lucky radical innovation. This occasional success cannot be repeated at will, but it may form the basis of success for many years and can be extended through continued improvement.

One strategy that can be employed has been termed the 'new game strategy'. This consists of using a particular technological strength of the firm to alter the main keys to success. The best-known examples are the recent advances in the efficiency of steel production in small mills, and a change in aluminium smelting technology which drastically reduced the electricity consumption of the process. This has given small steel producers an equal chance, and has removed the competitive advantage of cheap power from established aluminium smelters (Dussauge *et al.* 1992, 58–59).

Though a firm's strategic stance is determined by many factors, technology is one of the major ingredients that help to determine and maintain a strategic stance. Decisions on whether to pursue predatory expansion, or whether to seek export orientation and similar, have little to do with technology; but decisions about cost leadership, about technological pioneer status, about product differentiation and so forth are all given substance by technology. As technology forms a major weapon in seeking competitive advantage, a self-consistent strategy is bound to contain many technological elements. The next section deals with the formation of policy decisions that are the flesh and the bones of a strategy.

Public and commercial policy formation

Policy formation, whether in the public domain or in the commercial sphere, is an intensely political process. This means that policies are formed on the basis of value judgements and of attitudes, but it also means, first and foremost, that the acceptance of policy decisions is dependent on the relative power of the proponents of different sets of ideas. Strategic decisions often are the outcome of debates and power struggles. Technology assessment can help – only help – to increase the objectivity and rationality of policy decisions.[4]

We have stressed the similarities between technology assessment in the public domain and the information needs of strategic technology management in commercial firms. As technology assessment is an aid to policy formation in both the public and the commercial domain, looking at the differences and similarities between the two processes of policy formation may shed some light on the tasks of technology assessment in the commercial firm.

All policies are invariably formed with a goal in mind. All policy makers are guardians of certain interests and values and their aims depend upon these interests and values. If we regard the policy maker in the public domain as the guardian of the public interest – with the caveat that the public interest can never be homogeneous and that policy makers in power at best represent the interests of certain sections of the public – then in the realm of public technology policy their aims must be to foster and control technology in such ways as to

---

4 See the paper by Cabral-Cardoso (1996) for a discussion of the politics of project selection and the paper by Thomas (1996) on social and political processes in technology management.

obtain maximum benefit for society and cause least harm to it. What is regarded as maximum benefit, and how much should be done for the reduction of harm, and how much public money should be spent on technology, all these are questions that can be resolved only by political processes and can never be answered to the satisfaction of every member of society. Even policy makers are often divided on these issues and what policies emerge depends on compromise, deals and power struggles.

What governments try to achieve by supporting technology is to help the industry of their country to gain competitive advantage over industries in other countries, and thus to ensure successful trade and economic growth. Industrial success has, of course, many ingredients and it is a moot point whether government intervention in the technological aspects of industry makes a crucial difference. Most governments believe that it does, and there is universal agreement that, for example, Japanese government policy has played a major role in Japanese industrial success. Many ingredients that determine the technological prowess of a nation reside in the public domain: the basic research system; most of the educational and training system; parts of the system of technological research and development; the physical infrastructure (transport, power, communications, etc.);[5] government procurement; government information services; the legal and fiscal system; and, more elusive, the general cultural ambience. These hints will have to suffice, as our purpose is not to examine the technological potential of a nation, but merely technology policy proper. From our above list we select for further examination only the R&D system, the information system, and government procurement, as this is where public support for technology is given most directly.

We deal with support for R&D first. One aspect of this is support for fundamental research – a task that is financed almost exclusively by the public purse or by philanthropy. The amount to be spent and the topics to be researched are fiercely controversial. Though some of today's fundamental research may provide ideas for tomorrow's innovations, we shall concentrate on applied R&D, meaning R&D that is close to providing or supporting technological innovations or solving practical problems. No matter whether it

---

5 Some of the physical infrastructure is now often handed over to private enterprise, but is still subject to government regulation.

is carried out in publicly- or privately-owned laboratories or in universities, government will support some projects, within the limits of its budget, provided one of several conditions are met:

(a) The total cost of R&D, and the risk of never reaping substantial financial benefit from it, is so great that private enterprise will not invest sufficiently in it, yet the research is important either because of its future potential or because of its ability to solve a problem. Prime examples are nuclear fission and nuclear fusion research, research into the final disposal of nuclear waste, renewable energy research.

(b) The technology is important for the nation (or the world) but shows no potential for privately appropriable profits. Prime examples are research into health hazards and some medical research, environmental pollution, the greenhouse effect, road safety.

(c) Although the technology shows commercial potential, the cost of R&D is too great. In this case government will try to arrange a commercial partnership, but some government funding may still be forthcoming. Examples are found mostly in the aircraft industry.

(d) A technology is regarded as of overwhelming importance and the total national effort in it is deemed too small. In particular, the spread of this technological capability into small and medium sized firms is regarded as inadequate. The prime examples are microelectronics and information technology.

Government procurement is obviously important, as the armed forces, particularly, are voracious consumers of a never ending stream of new sophisticated equipment. It used to be argued that much military technological R&D diffused into civilian technological innovation. There is now much doubt about the validity of this argument and indeed the trend seems to have been reversed: the armed forces are supposed to benefit from civilian (or dual purpose) R&D (POST 1991). The US government used to rely heavily on procurement as a weapon for sponsoring technology. It was quite successful in the early days of integrated circuits. When these were too expensive for sane civilians to buy, the military bought them and thus brought in valuable cash that enabled further development, which eventually reduced prices dramatically.

Government information services, on the other hand, are regarded more and more highly. They consist of maintaining libraries and data bases, but also of facilitating the exchange of information among interested parties. In Britain, government is trying to knock heads together by encouraging discussion among industrialists and other experts to try and spot future winners (Cabinet Office 1993). The programme is known as 'Foresight'. The Japanese ministry for international trade (MITI) has long been famous for successfully coordinating industrial innovation in selected fields in the furtherance of the cause of Japan Inc. The US government supports cooperation among semiconductor manufacturers to retain US leadership in manufacture and development of logic circuits.

The formation of technology support policies is aided by information inputs which, even if they are not specifically termed technology assessments, closely correspond to TA. They are usually undertaken by civil servants or by consultants, sometimes by universities.

The other side of public domain technology policy is the control of technology. Control of technology occurs in many indirect ways, but the most direct means for the social control of technology is regulation. Regulation serves four basic purposes:

(a) The avoidance of danger to life and limb. Examples are building regulations, testing of drugs, safety regulations for motor vehicles.
(b) Ensuring that one user of technology does not interfere with other users. Examples are traffic regulations, rules for the suppression of radio interference, allocation of wavelengths for broadcasting.
(c) Ensuring minimum standards of comfort in the working and living environments. Examples are noise abatement, planning of land use, health and safety at work regulations.
(d) Safeguarding the natural environment. Examples are regulations for use of sprays, dealing with effluents and emissions, forestry regulations. Some environmental regulations overlap with those under item (c), but their main thrust is the guardianship of the planet earth for future generations.

The formation of regulatory policy in the public domain has no real counterpart in the private sphere, except that enlightened management might do more to safeguard the environment and the health, safety and well-being of its workers than is strictly required by statute and inspectorates. In the public domain, the basis of regulation must be research and information gathering, that is, technology assessment, followed by political processes of decision making. In the regulatory case, decisions are always hard fought. Nobody wants to be regulated, but not only are regulations unavoidable, they often offer commercial opportunities for new products or processes.

The avoidance of harm means controlling actual known risks, and regulating the use of technology in such ways as to cause minimum nuisance. Because a deteriorating environment not only causes a nuisance but poses severe long-term dangers, the protection of the environment must be a task for the public policy maker. This is particularly so because market forces are unable to protect the environment in anything like adequate ways (Braun and Wield, 1994; Braun 1995, 158–161 and 172–181).

The process of policy formation normally takes the following general route:

1 An issue is identified and is made part of the political agenda. This is done by institutions with watching briefs, by pressure groups, by the media.

2 Information is gathered and analysed, including the question of the need for action and an analysis of policy options. This is technology assessment.

3 The political process of decision making. This may involve a great deal of discussion and negotiation, it may involve parliament, or one or more quangos (quasi-non-governmental-organizations), or it may be resolved by a minister. The outcome may be laws, rules, regulations, tax concessions, financial loans or grants, the setting up of institutions, and so forth.

4 Finally, the policy is implemented. This is normally carried out by civil servants and special inspectorates, though the police, the courts, local authorities and other institutions may be involved.

The commercial analogy of the above steps is close.

1 An issue is identified and is made part of the management agenda. This is done by sales-staff, technology assessors, gate-keepers, the media, etc.

2 Information is gathered and analysed, including the question of the need for action and an analysis of policy options. This is

technology assessment, though it may trade under a different name, such as working party or policy group.

3   The political process of decision making. This may involve a great deal of discussion and negotiation, it may involve several departments, several managers, trade unionists or other workers' representatives, possibly the immediate neighbours of a plant, possibly suppliers or major customers, possibly government departments or planning authorities, perhaps the banks. Sometimes the decision may be taken by a senior manager without much ado. The outcome may be a change in products, a change in production facilities, a change in sales methods, an acquisition or sale, financial or organizational restructuring, improvements to sewage or emission treatment, etc.

4   Finally, the policy is implemented. This is carried out by line or strategic management, as the case may be.

As we have said, the public policy maker is the guardian of the public interest. The policy maker in the commercial firm, on the other hand, is the guardian of the firm's, or its owners', best interests. What the best interests are in detail is always debatable and is resolved in a kind of internal political process. In general terms, the most vital interest of the firm is survival, and that means sufficient profits to enable adequate investment to be made and to keep the owners (or shareholders) happy. The number of take-overs and mergers in recent years makes one wonder whether survival is indeed still the major aim of firms, but we must assume that, as far as the strategic technology manager is concerned, it is still long-term survival that counts. Just as national economies aim for growth, so commercial firms are not usually content with mere survival but regard growth as highly desirable, or even necessary.

The aim of the policy maker in the commercial firm is thus, in very general terms, reasonably clear, though the detail of policy aims may vary greatly from firm to firm, from manager to manager and from time to time. The technology policy within the firm must be subservient to the general aims and thus its task is well-defined: ensure that the firm has the technologies and products that will best ensure its long-term prosperity. Unfortunately, the time horizon of managers may not be as long as it ought to be and so-called short-termism, an enemy of strategy and of technology assessment, is rife.

The technology assessor must be aware of the interests he or she serves. Though the information provided should be objective and free of special pleading, it must be relevant to the aims of the policy maker. Though policy options are options to chose from, and the policy makers may not have clarified their aims before receiving the policy analysis, it helps if the assessors are aware of the general thrust of their policy aims. A good rule for technology assessors, partly safeguarded by the fact that they usually consist of a team, is: try not to allow your prejudices to influence your judgement and your analysis of policy options. On the other hand, there is no harm in declaring your values, though not to the exclusion of other people's values.

It is a point of much argument whether technology assessment should be carried out within the firm or by external consultants. It seems to us that an internal capability is very important and should form an integral part of strategic management. On the other hand, it can be very useful to employ external consultants to participate in, or even provide, individual assessments. We have stressed repeatedly that information has to be sought from wherever it resides, and consultants may be a good source of information and, sometimes, of wisdom.

What policy aims might the strategic management of technology pursue? We have discussed this matter under the heading of strategic planning and positioning, but it may be useful to provide a very brief summary.

The prime aim is to make a contribution to the profitability and competitiveness of the firm. This is achieved by making sure that manufacturing technologies are up to, or better, than the industry standard, that products are up to the required technical specifications, and that the firm is aware of technological and scientific developments that might offer future opportunities for better production technologies or new/improved products. Sometimes environmental or other regulations offer opportunities for the development of new products, e.g. automobile emission regulations led to the development of catalytic converters and electronic engine control systems. Infrastructural and ancillary technologies must be kept up to necessary standards. Similar considerations apply in many service organizations, transport undertakings, extraction and construction industries, and even in agriculture.

In commercial firms there is an aspect to technology policy that is somewhat analogous to the control aspects of technology policy in the public domain. The commercial firm must be interested in keeping its production technology as free from pollutants as it can. It must also be interested in avoiding all hazards to the health of its workers. The products should be as environmentally benign and as safe as possible.

The reason for all this is not just public spirit and altruism, admirable as these are. For several reasons, it is also hard-nosed good business practice. First, the firm must generally try to avoid conflict both with its workers and with the authorities. Should a manufacturing process prove dangerous, workers will protest against it and, sooner or later, the relevant factory inspectors will put an end to it. Should the process be polluting, neighbours may protest and, sooner or later, environmental inspectors will put an end to it. Hence the emphasis on preventing and controlling pollution and preventing health hazards in production.

Second, non-polluting production may actually save money, either because it simply is less wasteful, or because of the principle 'polluter must pay'. It is as well to remember the double meaning of the word waste: it is wasteful to produce waste. A process that avoids waste may well be the more economic way of production. It is almost certainly economic to reduce pollution because new legislation forces those who pollute to bear the costs of the necessary cleaning and disposal operations. Prevention is better than cure; avoidance is better than cleaning up. Desirable as it may seem to externalize the cost of pollution, increasingly tight regulations make this ever more difficult. In any case, carrying out the prevention of pollution and the cleaning up operations internally gives the firm control over this technology and its cost. This is preferable to bearing public costs beyond the firm's control.

Third, it is good for the image and reputation of the firm to be known to care about the environment. This is one aspect of what has been termed the ethical firm. To be ethical means to be honest, but it also means to care about employees, about the environment, and about general welfare. If the products of a particular manufacturer acquire the reputation of being safe, of saving energy, of being non-polluting and easily recycled, this must add to their appeal to the buying public. Indeed, many manufacturers have adopted the stance of the ethical firm and this is a welcome trend. Technology assessment can assist in informing the firm about dangers lurking in applications of technology and about new ways and means of avoiding pollution, of saving materials and energy, and, last but not least, about new opportunities of producing environmentally benign products in environmentally benign ways. The ethical firm must be well-informed. Unfortunately, this does not mean that the well-informed firm is bound to be ethical.

Information needs of strategic technology managers

The primary task of technology assessment in commercial firms is to satisfy the information needs of strategic technology management. The TA analysts – we shall call them analysts for short – have to seek out, marshal and analyse the information. Either the analysts themselves, or a different set of technology managers, depending on the particular organization, use the information to make analysed proposals for technology goals and for measures needed to attain them. Thus draft strategic technology plans are created for submission to top management. After possibly some iteration, top management will decide on a plan to be implemented by strategic technology management. The plan may be a routine annual one, or it may be the result of an ad hoc study, undertaken in response to some change: a new technological or commercial opportunity, a new danger, new management, a take-over or merger.

In all cases, the information required for planning purposes will fall into several categories:

- Are any of the firm's products or their technical features obsolescent?

There are several possible reasons for obsolescence. One is the so-called life-cycle of products. This affects demand for them, but more seriously affects profitability. As a new product comes on the market, it can command premium prices and if only one or very few manufacturers are able to produce it, they can obtain monopoly profits. As more and more manufacturers enter the market, price competition sets in, prices and profits fall. Eventually profitability will be very low. The product may continue at this level, or it may be overtaken by improved or entirely new products.

A product may become obsolete because of technological change, even when its life-cycle has not run its course. This may be caused by technological innovation or by changes in fashion, styling, legislation, or life-styles. Coal burning fires became obsolete when the clean air acts were passed. Radio valves became obsolete when the transistor and integrated circuit proved superior. Typewriters became obsolete when the word processor took over. Telephone dials became obsolete when everybody wanted press buttons. The wing mirror on cars became obsolete when side mirrors became fashionable. Galvanized steel buckets disappeared with the advent of plastic buckets. The long-playing record has been replaced by the compact disc. Plastic soles have largely replaced leather. The quartz watch has replaced the mechanical one. The list could be extended indefinitely. What is more difficult to know is the imminence or otherwise of the demise of a technology. Will the present-day television receiver be replaced by the digital receiver? Almost certainly, but how quickly?

- Are the firm's products up to the standards of the competition?

Analysis of competitors' products is a standard weapon in the competitive armoury. It is customary to draw up a list of features of a product from different manufacturers and compare these with the own product. It is also common practice to take competitors' products apart and investigate them thoroughly. If it turns out that, price for price, the own product is inferior, it obviously needs to be improved. If the own product is just different, market research may have to be carried out to see whether the difference is desirable. Cameras provide an interesting example of competition in features. These are relatively easily provided by electronics and are deemed necessary for competitive reasons. It is unfortunate that most of these extra features are pretty useless to the consumer.

- Are there new technologies on the horizon that might be suitable for the firm's range of products, either as new products or as improvements or supplements to existing ones?

The firm's own research laboratory may produce a stream of suggestions; the patent literature may provide many ideas; general knowledge, aided by consultants, may offer fruitful information, scientific and technical journals, fairs and trade exhibitions, data banks, are all sources of information. To guess accurately which of these many ideas will prove to be winners is an art that all managers would dearly like to possess. Only common sense, knowledge of markets, understanding of the technology involved, experience and consultation can help to provide the answers. And still, an element of luck is needed.

It is easy to provide any number of examples with hindsight. The disc-brake and the addition of electronic engine control systems improved the motor car; the condensing boiler is more efficient than the traditional domestic gas boiler, but is not catching on rapidly because of a price differential; the microwave oven has become widely accepted; the automatic teller has made huge inroads into retail banking.

- Are all the supply chains as good as they ought to be?

Do all the suppliers perform well, are all the make or buy decisions correct? Are there any major technological changes that the suppliers ought to incorporate? If a new product is contemplated, new decisions about what to make and what to buy have to be made and a new network of suppliers has to be built up.

- Are there any foreseeable changes in regulations, life-styles, fashions, which might offer opportunities for new products?

Even the highly regrettable increase in crime offers opportunities for new products, such as electronic burglar alarms or immobilizers for cars. Changed life-styles have brought with them elaborate equipment for gymnasia. Increased awareness of environmental issues and new regulations have brought about new opportunities for wall and roof insulation, domestic temperature controls, fire-proof furnishing fabrics. There is a generally recognized need for better batteries or fuel cells or for alternative fuels for cars. Though this has been known for a long time, it has proved elusive and the development costs are likely to be staggering. The future is still uncertain.

Concepts such as technology trajectory and techno-economic paradigm can give some help. It is obvious, for example, that in the present climate of opinion cars need not go faster, but need to have cleaner exhausts, use less fuel, and be safer. Though the murderous concept of the GT car is not dead, it is no longer dominant. It is equally clear that computer memories and the capacity of silicon chips will continue to improve, though nobody knows for certain whether there will be a radical change of technology. The present trajectory is unlikely to have run its course. Computer networks are here to stay and to expand, though there is plenty of scope for speculation on the relationship between e-mail and fax, for example. The latter may become obsolete because of rapid developments in Internet, World Wide Web and so on. What I am trying to say is that these theoretical concepts do provide some guidance, but they are very far from providing definitive answers.

There is both strength and weakness in sticking with the generally accepted and talked about active generic fields of endeavour:

information technology, bio-technology, new materials, energy technologies, environmental technologies (see e.g. Coates *et al.* 1994). Undoubtedly, this is where much of the action lies and one can glean something of the future by looking at developments in these fields. However, there is life in other spheres, and it may well pay to look away from the broad avenues into the small lanes of little explored fields.

- Is the firm's production technology as effective and efficient as that of its rivals?

The analysis of the performance of rival firms is standard practice. Indeed we often speak of industry standards, or best available practice, to describe what is normally achieved in the industry and what the maximum achievement is. These are useful yardsticks for measuring one's own performance. It is well known, for example, how many person-hours are needed in different car plants for the assembly of similar cars. The differences can be very striking. Whereas in 1989 the most productive car plant produced a vehicle in only 13.2 person-hours, the least productive plant needed 55.8 hours, roughly four times as long, for a comparable vehicle (Graves 1991, 267). It is often also known what the rate of rejects from a production line is and how efficient the quality controls are. The efficiency of just in time delivery systems and, hence, the reduced need for keeping stocks of parts is a further important parameter of efficiency of production. Does the production process (and the design of the product) make optimum use of energy and materials, is the layout of the factory the best possible, is the computer control as good as the latest reliable system? As in all aspects of technological innovation, it can be very rewarding to be a pioneer, but it is also rather risky. Hence the emphasis on the latest reliable system, rather than on the latest available one.

- Is new organizational knowledge forthcoming that might be useful to the firm?

Technology assessment does not concern itself with the overall organization of the firm, but is very much concerned with the organization of production. The efficiency of production depends not only on the hardware used, but is greatly dependent on how the hardware is organized into a production system.

- Are there new production processes on the horizon that might be suitable for the firm?

Almost everything that was said about technologies for new products is equally applicable to new production technologies. Many production systems have been completely revolutionized in recent years. Computer controls have been introduced, robots do much welding, paint-spraying, fettling, packaging and assembling. Total quality control, just in time delivery, the grouping of work-stations, the introduction of numerically controlled machine tools, computer integration, automated internal transport, are all recent examples of revolutionary change on the shop floor. In other industries, such as brewing, batch processing has been replaced by continuous flow processes. New steel-making and aluminium smelting have been mentioned. There is new textile machinery, new paper-making machines. New machines, equipment and organization for almost every industry.

- If new products are to be produced, what new production and process technologies will be needed?

Products and manufacturing technologies are inextricably linked. If it is planned to produce a new product, it is necessary to examine carefully what new manufacturing processes need to be mastered for its efficient production.

- How are the new product or production technologies to be acquired?

The immediate question is buy or make. Further questions are exact specifications and organizational requirements. A total quality control system, for example, does not only demand new technological arrangements, it mainly demands new organization, new attitudes and new skills. Some technologies may be difficult to obtain and perhaps some cooperation agreement with a rival firm needs to be arranged. If that is not possible, a way of bypassing the bottleneck needs to be found.

- What new skill requirements will arise out of the new technologies?

To operate and maintain new technologies successfully requires a great deal of training and/or the acquisition of new skills in the workforce. The identification and acquisition of the required skills and training form part and parcel of the strategy of introducing new technologies. Or will the new technologies cause de-skilling and, hence, labour conflict? Will they increase labour productivity to an extent that will necessitate the shedding of labour (down-sizing is the euphemism), because an increased volume of production cannot be absorbed by the market?

- What other requirements, such as R&D, maintenance, buildings, special supplies, will arise out of the use of the new technologies?

If a new product or production process is to be introduced, it is possible that it is not sufficiently developed and that a further R&D programme needs to be carried out. It is also possible that special building and services requirements might arise, or that new sources of supply for certain materials or parts need to be explored.

- Are present technologies environmentally as benign as they should be?

Are current production methods designed so as to eliminate as much waste as possible? In particular, does it eliminate or, at least, contain, hazardous emissions and effluents? A responsible manufacturer considers the product throughout its life. How much material and energy goes into its production, how much energy will it consume during use, how difficult will it be to recycle at the end of its life? These questions are important to society and the attitude manufacturers take may reflect upon their image, may affect sales of the product, and may save money.

- Are new regulations forthcoming that might affect some of the technologies in use?

Regulations undergo constant revision and it is well worth keeping ahead of them. Sometimes regulations require re-design of products, sometimes they eliminate certain products altogether, often they represent opportunities for new products.

If there is any new knowledge that might show some of the technologies in use in the firm to be health hazards, it may be worthwhile to eliminate these even before regulations force their elimination. Eliminating health hazards in advance of regulations is not only a wise precaution, it is also a way of avoiding conflict and improving labour relations.

- Are there any signs of shifts in public attitudes which might put into question some of the firm's technologies?

There are many examples of shifts in public tastes and attitudes. Energy saving has become a new concern, food additives are viewed with suspicion, synthetic materials are frowned upon in some contexts, leisure activities are booming, safety has become fashionable.

- Are there any government technology support measures available, or forthcoming, that might be of advantage to the firm?

Many governments support technological developments by a variety of measures, such as direct financial support, information, or training programmes. It is in the obvious interest of firms to take advantage of such programmes if they are relevant to the technologies they are trying to develop or to apply.

No doubt the list of questions could be extended, but what has been said should suffice to provide an overall view of what an industrial technology assessment might contain. It must be emphasized that assessments produced for different purposes, in different firms, about different technologies, will differ greatly. There is no one universal way. What is needed is wide-ranging consultation within and outside the firm, a wide-ranging search for information, and an open mind. The general methodology (STIP) can serve as a guide only. In further chapters we shall return to methodologies and to examples of assessments to complete the picture.

What technology assessment must do in all cases is to provide an analysis of impacts and consequences of all suggested policy options. It is these consequences that need to be analysed in as broad and long-range a fashion as possible – that is the true hallmark of technology assessment and distinguishes it from both technology forecasting and ordinary short-term policy analysis. The consequences may be commercial, environmental, organizational; they may affect management structures, employment, skills, safety; they may improve competitive position in some way but have costs in terms of lost loyalties or painful financial and personnel decisions. It all needs to be set out as objectively as possible, without fear of saying things that top managers may not wish to hear. If top managers cannot bear frankness and honesty, then they do not deserve to be at the top.

Before turning to these matters, we shall briefly explore, in the next chapter, the currently recognized dangers and worries associated with technology.

4

CONTEMPORARY PROBLEMS OF TECHNOLOGY

My purpose in reviewing contemporary problems associated with the use of technology is not to spread doom and gloom; my purpose is to show that these problems provide challenges and opportunities. Some may be solved by technological means, affording opportunities for innovative products and processes. Others may be solved by restraint and re-orientation in our uses of material goods, offering opportunities for new thinking and for creative action. Problems are here to be solved, not to defeat us. But to solve them, some cherished old patterns of thought and habit may have to be discarded.

I wish to alert the technology manager to these issues and to draw the attention of the technology assessor to them. An assessment that disregards problematic issues is not worthy of the name technology assessment. The technology manager must not use tactics reputed to be used by the ostrich; he or she must face reality, and indeed must look well beyond the conventional and venture into problematic territory.

It should become self-evident that each technology assessment must face the question: does this technology impinge in any way upon the problem areas of technology? If so, is the impact beneficial or, if it is not, can the technology be modified so as to make a more positive contribution to the solution of a societal problem? And if that is not possible, can the negative impact be ameliorated? Or should the technology be dropped?

Environmental issues

When speaking of problems associated with the use of technology, the first thing that springs to mind is the natural environment. This is not the place to rehearse all the arguments and repeat all the information available in numerous publications. Let it suffice to recapitulate very briefly what the main ingredients of the problem are:

1 Air pollution

(a) The so-called greenhouse effect, caused by carbon dioxide, the product of combustion of fossil fuels; by methane from escaped natural gas; and by propellant and refrigerant gases, the chlorofluorocarbons (CFCs). The great fear is that the increased concentration of these gases in the atmosphere may cause a rise in the average temperature of the earth, with somewhat unpredictable, but possibly catastrophic consequences for the climate and the distribution of precipitation.

(b) Destruction of the ozone layer. Some 25km above the earth there is a layer rich in ozone, and this helps to shield the earth from excessive ultraviolet radiation. This layer is being severely depleted by CFC gases and, despite some international efforts to reduce or eliminate the use of these gases, progress is not nearly as fast as it should be. One result is an increased incidence of virulent skin cancers and a new fear of the sun.

(c) A miscellany of effects caused by increased concentrations of gases such as nitrogen oxides, ozone in the lower atmosphere, sulphur dioxide (acid rain), carbon monoxide; and by the presence of pollutants such as unburnt particles of fuel (hydrocarbons) and soot. The effects range from the occurrence of smog, an increased incidence of bronchitis, asthma, hay fever and other ailments, to deleterious effects on trees and forests.

The chains of cause and effect are hard to establish and there is much conjecture involved in discussion of the effects of air pollution. The greenhouse effect and the depletion of the ozone layer, however, are pretty well established and may, ultimately, prove extremely serious problems for the flora and fauna of this planet, with possibly devastating effects on humankind.

2 Water pollution

Some of the air pollution eventually turns into water pollution, but the biggest effects are caused by two factors.

(a) Effluents from factories, farms and homes. These include high concentrations of heavy metals (such as lead or cadmium), nitrates, phosphates, a great variety of toxic chemicals, and all the other ingredients of a witch's brew that are not readily understood by any outsider. The effects, real or merely suspected, range from reduced male fertility, to Alzheimer's disease, to serious health risks to bottle-fed infants. Occasionally, dramatic spillages of toxic chemicals or oil occur and the results need not be conjectured, they are very obviously dead fish and birds in their thousands. A special case of water pollution is in coastal waters, where concentrations of toxic chemicals and bacteria from inadequately treated sewage can cause fish and shellfish, if eaten, to become acute, sometimes deadly, health hazards.

(b)   Fertilizers used in agriculture eventually get washed into rivers and lakes. The result is too many plant nutrients in the water, resulting in the growth of algae and a shortage of oxygen in the water (atrophication), causing fish to die. Other chemicals also eventually reach lakes, reservoirs, estuaries and coastal waters and can seriously affect not only aquatic life, but also the potability or other use of the water. The world-wide supply of clean water for drinking, cooking, washing and irrigation has become extremely problematic. Some people think that shortages of water will set the ultimate limits to human expansion. The only options are careful husbanding of resources, with good sewage treatment so that water can be re-used after primary consumption for secondary purposes, such as irrigation. Another option is the de-salination of sea water, but this makes huge demands on energy supplies, unless it is done by a clever combination of semi-permeable membranes and solar energy.

3   Miscellaneous pollution

The most obvious is trash. We are drowning in rubbish from discarded packaging, discarded batteries, discarded oil, discarded machinery, discarded everything. Growth in trash produced per head of population is relentless. This poses great problems of safe disposal, as some of the rubbish is highly toxic and polluting, and becomes hazardous when it seeps into the ground-water. It is becoming difficult to find space for landfill, and difficult to use old landfill sites for building or recreation because they contain so much toxic material. The apparently simple answer lies in recycling instead of throwing away, but there are serious technical and economic difficulties to be overcome. Some rubbish can be burnt and the heat used, sometimes methane can be extracted from rotting refuse. Some small beginnings with both the reduction and utilization of trash have been made, but much is left to be done.

There are relatively harmless nuisances, such as what has been termed noise pollution. Some of the noise pollution is caused by noisy vehicles and noisy machines, and is subject to amelioration by technical means. The noise from modern aircraft is slightly less deafening than that from older planes. Noise pollution is exacerbated by poor acoustic quality of buildings, and standards could be improved. Much nuisance is caused by wilful abuse: extremely noisy (or ever present) music and deliberately noisy motorcycles.

The mention of motorcycles brings us to traffic pollution. Not only is traffic the most prolific contributor to air pollution, traffic is also a major cause of death and injury, blighting many a young life. Traffic congestion is a sort of pollution in its own right that puts a major brake on our enjoyment of life in cities and, at weekends and holiday times, our enjoyment of the countryside. The solutions to these problems are obvious: less personal traffic and more and better public transport. Yet the will to implement these solutions is, mostly, lacking. There is also much that technology can do, but it needs to be very carefully thought out. The road to a traffic hell is paved with well-intended hare-brained schemes.

The deterioration of soil because of the use of heavy machinery, mono-cultures of crops, lack of organic fertilizers, lack of wind-breaks, and over-use of chemicals is also a serious problem. Soil becomes compacted and impoverished and substantial amounts get blown away by the wind. We can call this dust pollution.

Finally, we come to the destruction of the last remaining rain-forests. These harbour untold thousands of species of fauna, thousands of barely known plants, some of which might have medicinal and other uses. They are also home to people who have lived there since the dawn of humankind; and they act as important 'reverse lungs' that absorb carbon dioxide and produce oxygen. The use of technology has vastly accelerated forest clearance, yet there are no technical solutions to the problem of the disappearing forest. The only available solutions are political and economic.

We have not discussed the problems of exhaustion of natural resources, as these arguments are well rehearsed and it is obvious that the problems can be eased, or postponed, by technological solutions. More efficient use of resources, recycling, alternatives to rare resources, to mention just a few of the available means of amelioration. Slower innovation, allowing equipment to have a longer life, would also save resources, except when the new equipment causes substantial resource savings, for example when it is more energy-efficient.

When all is said and done, the sum total of these problems, combined with the relentless growth in the total human population that the planet is called upon to support, adds up to a real threat to the ecosystem. If we carry on regardless, we may be on the way to destroying the system that nurtures and sustains us. We have certainly destroyed, and are still destroying, countless species of fauna and flora, and yet we, as the dominant species on earth, ought to be the guardians of this planet and all life on it. Perhaps some environmental fears are exaggerated, but the sum total is serious, perhaps very serious, and if we are going to make mistakes, let us err on the side of caution. We in the industrialized world can easily afford the odd small sacrifice that may save the earth.

The impact on the environment of any technology being analysed must, as we have said repeatedly, form part of the assessment. There is much that technology can do to alleviate environmental problems. Hopefully, decision makers will welcome such positive impacts. There is also much that can be done to modify proposed or existing technologies to reduce their damaging effects on the environment and, yet again, it is hoped that decision makers will welcome such modifications. Finally, if a technology assessment shows damaging aspects of a technology, it is to be hoped that decision makers will either reject this technology or, at least, modify the damaging aspects, if at all possible. The technology assessment must set out clearly what damaging effects are well understood, which are subject to differences of opinion, and which are merely conjecture.

The omens are not favourable. Three separate reports, published recently by highly respected institutions,[1] are unanimous that the world's political leaders apply verbal 'greenwash', while continuing to act to the detriment of the environment. They pursue economic growth at any cost and are not prepared to grasp the nettle of environmental destruction. Indeed, they hand out enormous subsidies for activities that destroy the environment and, incidentally, increase poverty and destitution in the world. Damaging activities, such as intensive agriculture, over-fishing of the oceans, thermal power-stations, damaging industrial production, are all beneficiaries of government largesse. Alternative energy sources, on the other

---

1 John Vidal reported in the *Guardian Weekly* of 2 February 1997 on environmental reports published by the Worldwatch Institute in Washington, by the British government's own Panel on Sustainable Development and by the UN Environment Agency in Nairobi.

hand, receive very little attention. The situation in Africa is approaching catastrophe rapidly: water shortages, soil erosion, most species declining and many facing extinction.

Opportunities for socially useful technologies

There are very many technologies and, even more important, potential technological innovations, that can contribute to the amelioration of the problems listed above.

There is much scope for every type of cleaning, filtering and treatment technology for emissions and effluents. If small, efficient and cost effective water treatment plant were available, for example, then many a large consumer of water would be able to recycle water rather than discharging it into the sewer. Another example is oil filters for engines. If the filter is efficient in removing even the smallest particles from the engine oil, particularly in large diesel engines, the oil does not need to be changed at such frequent intervals (Rakos 1996). There are many possible improvements toward the reduction and disposal of rubbish. Manufacture for easy recycling is one answer, effective waste separation and recycling is another, clean burning of waste a third, and longer-lived products with less packaging yet a fourth.

To say that the technology for dealing with oil spills is in need of improvement is considerably understating the case. From time to time reports about promising technologies appear in the press, but where are they?

The greenhouse effect can be reduced if we reduce energy consumption and replace fossil fuels by renewable energy sources: biomass, wind, sun, waves. There are many possibilities, yet governments do little and private firms could do more. Nuclear energy is ideal from this point of view, but it causes so many other problems that it can hardly be considered a good solution to the energy question. Clean, safe, energy-efficient cars and good public transport should be priorities. Even energy saving in the home can be improved greatly: more efficient boilers, better temperature controls, better insulation, showers instead of baths, more efficient lighting.

Much remains to be done in medical technology: spare parts surgery – whether by mechanical means or, morally more problematic, with genetically modified animal organs. Artificial blood for transfusions might come about; diagnostic procedures and keyhole surgery might be further improved.

Agriculture could be improved if sewage sludge could be freed from heavy metals so as to become an effective organic fertilizer.

Traffic congestion might be eased not only by improved public transport, but also by home working, home shopping and so forth. Home shopping would have to be coupled with an efficient delivery system; home working may cause serious social problems as it isolates the worker, removes conviviality and makes it harder for workers to become effectively organized.

These are just a few examples, possibly obvious ones, but they may suffice to illustrate the point that problems are there to be solved, and that technology, coupled with the right organization and the political will, can solve many of them. I am sure that even relatively small entrepreneurs may hit upon good ideas that are both economically viable and environmentally favourable.

Technology, employment and skills

One of the perpetual arguments surrounding technology is its impact on employment. It dates back to the Luddites, and probably even earlier days. To strip the complex argument to its bare bones, we can put it like this: the immediate impact of production technologies is to save labour. Total employment in an economy will, however, be diminished only if the displaced labour cannot be deployed elsewhere. Thus the question is whether total demand in the economy can maintain full employment despite the labour saving effects of new production technologies.

The arguments attained new fury with the advent of computers and the new wave of automation they brought in their wake. In the past, labour displaced from agriculture found employment in industry, and labour displaced from industry found employment in administration and services. But the computer reduces demand for labour in these areas as well, and there is nowhere for displaced labour to go. The only solution is a continuous increase in total consumption of goods and services, sufficient to absorb the increases in labour productivity in all sectors. Another solution is to distribute what employment remains available by reducing working hours and working years without loss of income. For loss of income means loss of purchasing power and this exacerbates the problem of employment.

To increase consumption of material goods is a hazardous undertaking, because of its impacts upon the environment and upon depleting natural resources. To increase demand for services is easier, except for two problems. One is that some services, such as

99

tourism, also have a devastating impact upon the environment. It causes congestion on roads and in the air, makes huge demands on energy supplies and on water and sewage services, and degrades previously unspoilt sites. It is also somewhat self-defeating, as desirable holiday destinations become undesirable with the growing numbers of people who visit them.

The second problem with increasing demand for services is their financing. There is potential demand for almost endless recreational services, medical services, public transport, better housing, better educational facilities, cleaner cities, more cultural offerings, better care for the elderly, and so forth. But how do we pay for these? Increasing taxation is resisted and other means of financing some of the desirable services are proving difficult to envisage. Those services that can be provided by private enterprise, and it is by no means all of the desirable services, are provided selectively for those who can pay for them.

Regrettably, the post-war period of full employment has come to an end. All countries, though some more than others, now have substantial unemployment and under-employment. In some countries, much work is extremely badly paid and there is unhealthy competition between low-paid jobs and social security payments. Most people are insecure in whatever employment they find; many who seek full-time employment have to settle for part-time work. Other workers work extremely long hours; the trend to part-time work and part-time unemployment has not brought with it a universal reduction in working hours and the fair sharing out of what employment there is. Unemployment hits the young particularly hard and all the training and education in the world cannot solve the problem unless demand for trained workers begins to match supply. We shall not recapitulate the tales of woe that are told in statistics covering crime, drug abuse, and poverty, as these are political, not managerial, matters. The employment problem, though, is at least in part a managerial problem.

Nor shall we rehearse the arguments about unsatisfied demands for all goods and services in developing countries. These too are problems beyond the potential for action of the manager. Rules of international trade expressly forbid discrimination against goods produced under extreme hardship (e.g. child labour) or with extreme disregard for the environment. There is nothing in these rules that forbids the individual buyer of such goods to ask whether they are acceptable or whether a less shady source of supply ought to be found.

There are numerous studies on the effect of information technology on employment. It seems to us, however, that the question put in this way is unanswerable. It is almost pure conjecture to try and find the balance between the numbers of workers displaced by information technology, against those employed in producing goods and services to satisfy new demands created by information technology. We can only view the overall picture of increased unemployment, but we cannot even be sure how much of this is caused by macro-economic and political circumstances in the world, and how much by the world-wide application of computer and information technology. Or how much is caused by the great redistribution of income from the poor to the rich that has taken place in some developed countries, and between developed and developing countries, in recent years.

Skills are a somewhat different story. Demand for purely manual skills has certainly decreased in consequence of using sophisticated automated machinery. On the other hand, new demands for multi-skilled workers have arisen. For the sake of flexibility, workers have to be deployed wherever a momentary need arises, and the more jobs they can tackle, the more useful they are. It is also true to say that sophisticated computer-controlled work-stations (formerly machine tools) still operate at their most efficient when in the hands of old-fashioned skilled workers. There are also new demands for skilled maintenance workers, as modern machinery is not as totally reliable and as free of maintenance as one might think. Computer and communications experts are in demand, though in computer programming the competition for jobs is world-wide. Total employment in manufacturing industry has declined since its peak, but some of the statistics are misleading because so much work that used to be tackled by industrial employees is now done by outside contractors, who are part of the service sector. Transport, catering, cleaning, building maintenance, even aspects of design and R&D are now often bought in rather than produced in-house. The reasons for this are cost savings and increased flexibility. Outside workers often get worse conditions and lower pay and their services can be called upon, or dispensed with, as needs ebb and flow.

Have computers led to de-skilling? To some extent, the employment of skills is negotiable; machines can be operated with more, or with fewer, skilled workers (Wilkinson 1983). But perhaps this is becoming less and less the case, as truly computer-integrated manufacturing systems come on stream. On the whole, it is probably true to say that demand for manual skills has decreased, but demand for more abstract skills has risen. Who can tell whether the sum total of skills in demand has increased or decreased? Much of the answer to this question depends on statistical definitions and classifications.

Why and how does all this concern the technology assessor or technology manager? Clearly, it is not in his or her power to alter macro-economic or political circumstances. Neither can managers increase employment in their own plants, unless this can be economically justified. They need not, however, view each job done by humans as a candidate for automation. Though it is true that robots do not take tea-breaks and do not make wage demands, it is also true that they cannot cope with unforeseen situations. Managers are able to make important contributions towards improving employment prospects. They can value their employees as assets that are to be developed and to be cherished. They can use the knowledge, the creativity, and the initiative of their workers to improve productivity and quality. They can deliberately improve working conditions and encourage the use of skills, rather than attempt to wrest all control out of workers' hands and use them only as living machines and candidates for elimination. They can test all proposals for changes in production that reduce the demand for workers on their true merits. When all things are considered, this kind of rationalization is not always rational or inevitable. Sometimes an apparently rational case is made out on the basis of drawing the boundary of the analysed system too narrowly and not considering secondary effects and consequential costs. All in all, managers ought to regard their workers as valuable human beings, not as nuisances that automation has not yet managed to get rid of.

Globalization

Globalization has become a buzz-word and a subject of much concern and study.[2] But what does it mean and why is it a problem? It means that capital has become internationally mobile. More important to us, it means that many large corporations now operate internationally and that they can switch their production from country to country. In this way they can force down wages, both by actually operating in low-wage countries with low wage-overheads

---

2 For thorough reviews of themes connected with globalization see Freeman and Hagedoorn (1992) on globalization of technology, and Jahoda (1992) on social and political issues, both published by the CEC.

and by threatening to do so. By the same token, they can enforce labour discipline and meek acceptance of worsening conditions. They can bypass many government actions, particularly regulations and taxation.

On the other hand, most countries – and regions within countries – compete for what has become known as inward investment, meaning foreign investment flowing into the country concerned. This competition can give the investing company real advantages: investment support by direct grants, provision of infrastructure, tax and other concessions. The art of selecting a location for building new plants has become quite sophisticated. Before investing, major investigations are undertaken, covering aspects such as government incentives, quality of life, cost, skills and docility of labour, political and financial stability, taxation, regulations, membership of supranational bodies (such as the European Union), accessibility and general infrastructure, local markets, availability of supplies. This type of investigation is a form of technology assessment and we shall return to it in Chapter 6.

By and large, international corporations are very much based in one country and workers and managers in other countries are not always treated as full equals. This removes some decisions and some of the more glamorous aspects of the firm's activities from their grasp. No multi-national corporation has more than half its investment outside the mother country (most have considerably less) and the culture of the mother country strongly influences the way the corporation operates.

Globalization presents major challenges to management. The activities of widely dispersed units have to be coordinated to make use of the advantages of each site, and to meet the competition in each location. Some firms produce the same product in several locations (viz. the world-car), others specialize in certain aspects of production at each site. If a firm locates all the activities relating to a particular product in a single site, or closely related sites, this is known as 'global focusing'. If a company spreads all activities relating to a product across the globe, yet manages to coordinate them, this is known as 'global switching'. All global firms need an elaborate network of communication, an elaborate system of coordination, and an equally elaborate system of control. 'The three strategic phenomena – global networking, global switching, and global focusing – all seek to address these coordination aspects within the context of timely and effective management' (Howells and Wood 1993, 142–152).

The ascendancy of globalization is impressive, both in terms of world-wide employment by non-domestic companies and in terms of inward investment. In the automotive industry about 30 per cent of the workforce is employed by non-domestic companies; in chemicals the proportion is about 55 per cent. During the years 1980–1988, some US$250 billion were invested by foreign firms in the United States, while some US$150 billion were invested by US firms abroad (Howells and Wood 1993, 53, 70). This does not mean, however, that more and more factories are springing up all over the world. Indeed many of the global firms, though spreading their activities over the world, rationalize by concentrating their activities on fewer sites.

One of the aspects of globalization that worries some people is that R&D support schemes in one country may in fact mainly benefit countries other than the donor country, if the know-how gained through the scheme is used elsewhere by the global firm. This is only one aspect of the general problem that much government activity can be bypassed or nullified by global firms. As far as the technology assessor is concerned, these matters are something to be aware of and to consider in particular cases.

The aspect of globalization that influences managerial decisions most strongly is increased competition. This is not necessarily the result of activities by multi-national corporations, it is the result of the spread of technological capabilities to more and more countries and of free trade. Much is made of free trade as an agent of economic growth, but it is also an agent of increasingly vicious competition. Particularly worrying is that international trade agreements specifically forbid discrimination against goods on grounds of the way they are produced. Thus child labour, environmental abuse, and many other ills of some industrial and semi-industrial production are protected against action by consumers in advanced countries. Not only can advanced economies not hope to compete with developing countries in the production of undifferentiated staple products, they are powerless to protect the most oppressed workers and the most abused environments. Environmental problems are global, however, and eventually may catch up with us. Technology assessors should be aware of these issues and point them out, possibly suggesting defensive action, wherever applicable. This is not to say that employment in developing countries should be put in jeopardy – they need it badly – but that action against abuse of workers and against environmental degradation should be taken whenever possible.

The speed and direction of innovation

If competitive pressures force firms to innovate too fast, several things can, and do, happen:

1   Production technology is not given sufficient time to pay for itself – it has to be written off faster than is desirable and long before it is physically decrepit. This adds to costs. In the case of production machinery, these costs may, or may not, be partly or wholly balanced by the savings made through using more efficient machinery of recent vintage. In the case of product innovation, the cost might be balanced by increased profits on a new product. In many cases, the firm is likely to lose and the consumer certainly pays more than if the pace of change were more sedate.

2   Because machinery and products are changed so frequently, the learning curve is never given a chance to run its full course. That means that productivity never reaches its potential optimum, and indeed runs well below it for some considerable part of the life-cycle of a product and/or process. One aspect of not reaching the optimum on the learning curve is that managers and workers never reach a stage when they are at ease with their work. There is constant tension and frequent crises.

3   The cost of R&D is increased because of the pressure of time, and many teething troubles are not eradicated by the time the product or process reaches the market. Being in a hurry with R&D may mean cutting corners, it may also mean that the work is not carried out in a logical sequence – each step waiting for the outcome of the previous step – and thus more blind alleys might be explored than would otherwise be necessary.

4   Waste is increased as obsolete goods are disposed of.

5   Workers and consumers become disorientated by too much novelty, especially as the mania for innovation spreads into all spheres of life.

What are the conclusions for technology assessors and managers? First, they must ask themselves whether the proposed innovation will not only create a new product, but whether it will be a truly better product. Innovation for its own sake is pointless, and though advertisers make much of the adjective 'new', the public has become, rightly, somewhat suspicious of it. An innovation can only be justified if it produces a truly improved product, or a product that serves a real, hitherto unsatisfied, need. Even if the pressure of competition makes an innovation unavoidable, the above two criteria should still be borne in mind – though what anybody regards as a truly improved product is a matter of opinion. I would regard a safer car as a better car, but would not consider a faster car to be better. I would consider the development of artificial spare parts for humans as a good technology, whereas I am extremely indifferent to the breathtaking rate of change in computing. I would regard the development of renewable sources of energy as a high priority task, whereas I am indifferent to the provision of ever more television channels that produce the same dreadful programmes and even more dreadful advertising. You may not share my tastes, but you will see what I mean when I say that what constitutes technological improvements is a matter of opinion, though some more objective criteria of real need and social utility can be employed.

What is a real need? The question is a complex one and the answers given depend more on systems of belief than on objective knowledge. I have no intention of exploring this issue, except to state my belief bluntly: I believe that most private needs for goods and services can be satisfied by existing products and I do not believe that the fact that novel goods or services are bought proves that there is a real need for them. On the other hand, I believe that very many new goods and services are required and are not produced because there is no effective demand (demand backed by cash) for them. This is either because they are public needs and the public purse is empty (or misapplied); or because they are needs of groups that cannot pay for their satisfaction – the poor, the weak, the disabled, the fauna and the flora and the rest of the natural environment.

> What society really needs are technologies and attitudes that help to solve, or alleviate, societary problems. What we do not need are technologies designed merely to stimulate flagging demand by making products of technology subject to fashion and by producing ever new toys and gadgets.
>
> (Braun 1995, 185)

True needs that remain unfulfilled are those of weak individuals or groups and those that serve the common good or interests that have only proxy representation. The needs of the underprivileged are obvious and we need not explore them here; just remind managers that they exist. The groups that have no direct representation are mainly wild plants, wild animals and open spaces, whose welfare is the concern of only a few voluntary organizations. The public good that we are most concerned with, and that technology has done most to damage and can do most to salvage, is the environment. Some things can be done for the environment even by private initiative and within the confines of market considerations. Other things cannot – they must be the concern of government.

Much energy-saving innovation and innovation that improves safety can be produced by relying on market forces. With the right pricing policies, the following energy-saving products might prove sufficiently attractive to make their way in the market: energy-saving domestic and industrial heating; energy-efficient refrigeration plant; energy-efficient cars; easily recycled goods, provided the consumer were to pay for the disposal costs of all goods; goods free of toxic substances, provided they were adequately labelled; fire-resistant fabrics; and many many more. But it is doubtful whether effective road traffic management systems (and we do not mean road pricing, but safety management) can be introduced by anybody other than public authorities. We cannot see private enterprise coping with oil spills, though improved equipment for cleaning up the mess might, and should, be developed by commercial firms.

Technology managers clearly cannot solve problems that only government can deal with. On the other hand, they can contribute to making socially desirable marketable technologies available. In this way they can make a direct positive contribution, and this might gradually shift public opinion toward public solutions to environmental and social problems.

5

METHODS USEFUL IN
TECHNOLOGY ASSESSMENT

Technology Assessment may be regarded as a very general method of looking at the likely broad consequences of decisions on technology. Perhaps TA is more an attitude than a method – the attitude of attempting to take a holistic view of technology within its broad social setting. Technology assessment is a means of avoiding too narrow a view – tunnel vision – and too short a time horizon – myopia. People looking at their very own technological innovation tend not to observe the environment in which it exists, and tend to see only tomorrow and forget the more distant future. Taking a narrow and short-sighted view makes people prone to committing grave errors of judgement.

TA does, of course, use a variety of methods, but these are largely borrowed from the various social sciences. There is no magic in methods – though they may carry high-sounding names and be described in esoteric language, the essence of useful methods must be the systematic application of common sense and logic. If a method does not make sense to you, do not use it. Do not put your trust in magic formulae – they cannot overcome the basic uncertainties and imponderables of gazing into the future. Having said that, there is nonetheless a number of methods – or, more appropriately, approaches – that can usefully be employed in TA and we shall discuss some of these in the next sections. The main strength of methods is that they help the assessor to be systematic and indeed the main feature of TA methodology is that it helps to organize knowledge in a useful manner.

Principles of forecasting

Technology assessment aims to assess the likely consequences of the introduction of a certain technology. As the consequences have not happened at the time of the assessment, and even the technology may not exist, we are really trying to foresee the future, both of the technology and of its effects. Thus an important aspect of TA is forecasting, both technology forecasting and forecasting of the environment – in the broadest sense – into which the technology will be introduced and that will interact with it.

The quest for knowledge of the future is an irreducible aspect of planning actions and events that will occur in the future. Yet the only certain knowledge that we can have about the future is that we cannot know it. Thus we can dispose of the first fallacy about forecasting: it is not an attempt to know, or even predict, the future. Prediction falls into the realm of prophecy and is not subject to the rules of either logic or of science. Each individual is free to chose which, if any, prophecy to believe in; but rational individuals will not base their actions on such beliefs. Having said that, there are, of course, some future events that can be known. We know with a great deal of certainty how the sun and the planets will behave for many centuries to come, we know the passage of the seasons, we know that all living creatures will age and die. These are examples of well-documented natural events that are completely predictable, yet even then some details remain uncertain. Though we know that spring will arrive on 21 March of every year, we do not know how warm, wet, or otherwise it will be. Neither do we know whether, and when, apparently fixed events determined by humans might be changed. When we look at the longer term, we cannot be sure that court sessions will not be changed or abolished; even the calendar may change. We know that the future of political, social, human and economic trends and events is inevitably highly uncertain. But, and that is the point of forecasting, we also know that our actions shape the future. *Forecasting is an attempt to glean what effect different present actions and decisions might have upon the future.* We try to gaze into the future in order to learn how to shape it in desirable ways.

We do know that technology will change and develop, but we do not know either the pace or the details of such developments. But saying we cannot know, though true, is not good enough. Our actions do influence, and even create, a future; hence we are bound to ask what this influence is likely to be. If we are trying to introduce a new technology, we are bound to ask how likely it is to be successful in the market place and how it will interact with the world into which it will be introduced. How do we reconcile the paradox of our inability to know the future and our need to foresee the consequences of our actions upon the future?

In essence, we try to imagine different plausible futures and to assess the interaction of our technology with its environment in each case. We imagine different plausible changes in the social and natural environment, as well as different development paths for the technology we are introducing. The problem of forecasting is not to know the future, but to obtain a range of likely future situations. This gives us some guidance on what actions we need to take at the present and how we might adjust our actions to the unfolding future situation. We may reach a conclusion on what our best action might be under the circumstances that are, in our opinion, most likely to prevail, but we have to watch for certain events to either confirm this action, or modify it in the light of such events.

> futurists try to define a range of alternative futures and to use that full range of alternatives as the basis for planning. . . . Surely the most important measure of a good forecast is not whether it is right or wrong, but whether it pushes development in a useful direction.
>
> (Coates *et al.* 1994, 27)

This is begging the question of what a useful direction might be, but in our context useful is likely to be interpreted as useful to the firm, hopefully in an enlightened way.

Forecasting is the art of postulating a range of likely futures, enabling us to make reasonably informed choices of actions affecting the future. What we regard as plausible is invariably strongly influenced by the present we know, though it may also be influenced by wishful (or fearful) thinking. The term scenarios is often employed to describe what the forecaster does: invent a range of plausible trends, situations and events, influenced and modified by a plausible range of parameters. We may, for example, describe the future mix and total use of energy sources. The parameters likely to modify this are total GNP, cost and mix of primary energy, technological developments in energy supply industries, developments in transportation, developments in building technology, environmental concerns and legislation.

The literature, as is the wont of scientific literature, is full of definitions and typologies of different kinds of forecasts. None of this is of much interest to us, as the only type of forecasting that TA is concerned with falls into the category of normative forecasting. The question we ask is: what do we need to do in order to achieve the most desirable results. We do not have a grand scheme of modifying the future – as might be the purpose of normative forecasting on a grand scale – but we do wish to make the best of our technology in the future. This amounts to a modicum of deliberately changing the world, and hence forecasting for the purposes of TA is normative forecasting, albeit on a modest scale.

Methods of forecasting

In principle, we may distinguish three basic methods, though in practice there are many variations and hybrids available. A well-known survey by Jantsch in 1967 distinguished roughly a hundred different forecasting techniques (Encel *et al.* 1975, 65).

Extrapolation

The method of extrapolation essentially consists of obtaining historical data and fitting a curve to them which extends into the future. The fundamental assumption of extrapolation is that things will continue as before. Thus, if the number of airline passengers has increased according to some mathematical formula (curve) for each of the past many years, we assume that it will continue to increase along the same curve (in accord with the same formula) for many more years. It is very important in this type of forecast to obtain a good set of historical data in order to obtain a good fit of an extrapolated curve, and thus an accurate mathematical formula for the dependence upon time of whatever variable we are studying. Clearly, the historical data must cover a sufficient period to give us some confidence in our equation, and the future horizon must not be long compared with the time covered by the historical data. There would not be much confidence in extrapolating over twenty years into the future on the basis of data collected for a mere five years. The periods covered by both history and forecast must be clearly stated for any forecast to be worthy of the name.

The main pitfall of extrapolation is an unforeseen change of trend. Circumstances that caused a certain trend to prevail may change, either subtly or radically, causing the trend to change, slightly or greatly. The best-known example is the incorrect extrapolation of trends caused by not considering either saturation or substitution effects. The growth in sales of certain goods may appear to follow a linear, a quadratic, or even an exponential curve for a while, but before long the market will saturate, or the items in question will go out of fashion and be displaced by some other items. The result is some form of S-curve. This is well-known and it is very easy to forecast an S-curve; what is more difficult, and requires observational subtlety, is to forecast at what time the S-curve will approach its asymptotic value and what this value will be.

In the case of a simple full substitution of one product by another, we obtain a so-called logistic curve

$$f = 1/[1 + \exp b(t - t_0)] \qquad \{5.1\}$$

Where f is the proportion of the market captured by the new product and t is time. $b < 0$ is a constant and t_0 is the time at which sales of the new product equal sales of the old product ($f = 0.5$). At large t, the new product tends to replace the old one entirely and f becomes 1. The assumption that leads to this equation is that the rate of adoption of the new product is proportional to the fraction of the old product still in use.

This substitution curve is a special case of a more general logistic curve

$$y = K_1 + K_2/[1 + \exp(a + bx)] \qquad \{5.2\}$$

where $b < 0$, y tends to $K_1 + K_2$ as x tends to infinity, and y tends to K_1 as x tends to minus infinity. At $x = 0$, $y = K1 + K2/[1 + \exp(a)]$.

An example of a substitution curve (5.1) with $b = -0.5$ and $t_0 = 5$ is shown in Figure 5.1.

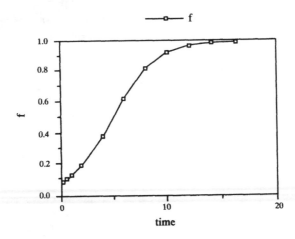

Figure 5.1   Substitution of a product by a new one as a function of time. This is an example of a logistic curve (or S-curve).

S-curves can describe all kinds of growth phenomena. If, for example, we plot R&D effort against some measure of technological progress in a well-established technology, measured in gates per silicon chip, or as engine efficiency and such like, we obtain an S-curve showing the growth of effort needed to obtain a 'given amount' of progress. We have referred to this phenomenon as the law of diminishing returns in R&D. To draw an S-curve for more than just demonstration purposes, we need empirical data to obtain values for the constants, and even then we cannot be sure that circumstances will not change in such a way as to invalidate the past empirical data for the future.

If a change in trend is caused by an unforeseen event, the whole method of extrapolation can be thrown into disarray. The unforeseen event may be a sudden rise in the price of crude oil, such as occurred in 1973; or the sudden realization that a commonly used substance poses a major hazard to health, such as was the case with asbestos; or political events such as the Iraqi invasion of Kuwait in 1990. It may also be a radical technological innovation, such as the transistor or the jet engine, which makes the observed technology suddenly obsolete. Or it may be a change in regulations, such as the clean air act that made the domestic use of coal obsolete in British cities. If a method of extrapolation is to be credible, it must state clearly that it either foresees no change in material circumstances or else it must postulate changes and show how these might affect the trend. In this latter case the extrapolation becomes not a single set of figures, but a range of plausible alternative trends correlated with certain extraneous circumstances. We might say, for example, that if GDP in a certain country over the next five years were to rise by an annual average of 2 per cent, inflation were to remain at 4 per cent average and the price of motor fuel were not to rise above inflation, then the total number of kilometres travelled by all cars would probably be X million per annum. We might further attempt to quantify – from historical data and plausible assumptions – how X would change to Y if the parameters changed in certain ways. In economic language, we need to know the price elasticities of car travel, i.e. the change in kilometres travelled caused by a given change in the relative price of motor fuel and changes in disposable income. Such scenarios can be useful if we are planning, say, the refining capacity of a petrochemical plant or if we are planning to introduce a new, more frugal, engine onto the market.

Trend extrapolation can give us a variety of scenarios if we look into the causes of a certain trend and postulate possible changes in these causes. Each change leads to a different extrapolation and thus we obtain what forecasting is all about, a statement of the kind: if variable x changes over time in certain ways, then variable y is likely to change in a predictable manner.

In summary, trend extrapolation can give useful results, provided we have good historical data and provided we take an intelligent look at the causes of the trend and the effect of possible changes in circumstances.

We should briefly mention one method, borrowed from economics, the so-called input–output analysis. Economists prepare input–output tables that show the flows of trade between sectors of the economy. The sectors are usually aggregated in some standard way according to the International Standard Industrial Classification (ISIC) and the tables are produced at fairly long intervals (usually ten years) and are several years out of date by the time they are published, so that they give a historical picture.

If there is substantial technological change in one sector, this may change both the inputs and the outputs of this sector. In other words, the sector may require different supplies and may need to find different customers. By looking at the pattern of trade of the sector undergoing change, we can assess how this pattern will change in time. We can study the effects of substitution or product innovation on intersectoral trade and, thus, on growth or shrinkage of sectors. For example, the motor industry has in recent years become a substantial customer of the plastics, electronics and aluminium industries. Plastics have replaced leather; aluminium and plastics have replaced steel. Electronics is not a substitute but a new product for the motor industry. If we analyse the input–output tables with technological trends in mind, many of the new trade patterns can be foreseen. This allows us to draw conclusions about some industrial developments.

Input–output tables are used to forecast the effect of investments in certain industrial sectors upon other sectors. If, however, major technological changes occur in the investing sector, then its pattern of expenditures will change. Similarly, if technological changes cause labour productivity to grow in certain sectors, then trade flows shown in input–output tables will have an altered impact upon employment.

It is now customary to look closely at what is known as supply chains. These may be regarded as the microcosm of input–output tables. They have the advantage of being up-to-date and looking at a specific industry instead of at a statistical conglomerate. Changes in technology affect supply chains and foreseeing and planning these changes may be an important aspect of any forecast undertaken as part of a technology assessment.

Expert opinion

One of the best-known methods of forecasting is the so-called Delphi method. In essence, it attempts to obtain a consensus of opinion among experts in a certain field on the likely future in their field of expertise. The Delphi method is most commonly used if the question is not so much whether a certain event will occur, but when it will occur. This leaves open the possibility of saying 'never'. Questions posed are usually of the kind 'when will a reliable and effective artificial heart (or heart from genetically modified animals) be available for transplantation into humans?'.

Usually the Delphi method involves more than one round of questions. The answers in the first round are analysed and the analysis is submitted to the same experts in an attempt to find consensus. Thus the second round might approach the experts with a statement something like the following: '60 per cent of the experts suggested that the event in question will occur within ten years. In the light of this majority opinion, would you like to modify your original estimate?'

The Delphi method is based on two articles of faith. First, that experts working in a field of science or technology have a good feel of how the field might progress and when certain key results might be obtained. Second, that the consensus opinion of several experts is more reliable than the opinions of single experts. In the light of past experience, it seems that these two assumptions are often, but by no means always, justified. As long as the time horizon is reasonable and no unforeseen events occur, experts do know what to expect within their field. What is a reasonable time horizon is a matter of opinion. I believe that anything further than twenty years into the future is entirely unforeseeable and forecasting beyond that time horizon should be attempted only under exceptional circumstances. The Delphi method suffers from the problem of finding it difficult to create scenarios, though it is possible to use its results as inputs for scenarios created by the team of technology assessors. In principle, it is also possible to ask experts a whole range of questions about what they think might happen if external circumstances changed. This can create something like a scenario of events under changing circumstances, but it is difficult to obtain and maintain the interest and cooperation of a sufficient number of experts for a complex and lengthy series of questions.

In Japan a major Delphi survey in the field of science and technology is conducted every five years. It attempts to forecast scientific and technological developments by asking experts to state the time at which certain technologies, or certain knowledge, will have become available. The fifth Technology Forecast Survey included 1,150 survey topics, ranging over a huge field from energy and materials technologies to medicine and life-styles. There were nearly 2,400 respondents. The time horizon of the forecast was thirty years and thus the most difficult technologies, such as a practical fusion reactor[1] were predicted to happen at that time. Many experts expected it to take much longer and, in my view, putting the time even at thirty years means that the event is not at present foreseeable. The mean of the expected time at which events were forecast to take place was, for most topics, about ten to twenty years. Interestingly, the development of effective methods for the prevention of Alzheimer's disease was expected, on average, to occur in twenty years. We shall quote just one question from the field of information and electronics to show the form this survey took. 'Practical use of technology easily enabling processing of patterns with line spacing down to 10 nm'. The mean time at which this was expected by the experts to happen was the year 2003 (NISTEP 1992, 6).

Expert opinion can be used without recourse to the formalized structure of a Delphi study. There are several ways of doing this. The simplest is to send out a questionnaire, similar to a Delphi questionnaire, but without going to a second round. A preferable, but much more labour-intensive method, is for the analysts to conduct interviews with selected experts. These can be additional to a questionnaire in order to elucidate the thinking behind the answers given, but often comes in place of a questionnaire. The interview can be fully structured, meaning that all the questions to be asked are prepared in advance in the form of a questionnaire. Probably a better method is the semi-structured interview, where questions are prepared in outline, but the interviewer remains flexible and allows the interlocutor to express views and opinions freely, possibly interspersed with anecdotes. As long as the interview does

---

1   A fusion reactor should produce energy on the same principle as that underlying the energy generation of the sun: by fusion of hydrogen nuclei into a helium nucleus.

not become too rambling and does not stray too far from the topic, this method can be very illuminating and bring answers to questions that should have been asked, as well as to those that were asked. The above methods can clearly be used in attempts to forecast, but are equally useful in attempts to elucidate opinions about directions of technological development and about attitudes and interests in relation to the application of certain technologies.

Difficulties arise because experts are busy people and are often unwilling to give much of their time. A more serious difficulty, especially in the commercial world, lies in the fact that both TA analysts and experts like to play their cards close to their chest. Secrecy is the enemy of broad-based knowledge, but is clearly not entirely avoidable in the commercial context.

Apart from the person-to-person interview, there are collective techniques that are often used, not only for forecasting but more often for elucidating opinions, attitudes and, very often, for producing ideas or clarifying and agreeing plans. Among the better-known techniques is brain-storming, where a group of experts get together and express their views without fear of making fools of themselves. The point is that they are encouraged to express even wild ideas in the hope that something constructive and reasonable will come out in the end through a process of constructive criticism. Another method is the workshop, essentially the same, but without encouragement to float wild ideas. It is a more restrained and more planned exercise, where participants are encouraged to express their thoughts frankly and without too much caution, but not without self-critical constraint and previous preparation. Both brain-storming and workshops depend critically upon good chairing. Participants must be encouraged to speak their minds, but must be discouraged from rambling on aimlessly and endlessly, so that every participant gets a chance to express his/her thoughts.

The methods of questionnaires, interviews, brain-storming and workshops are extensively used in TA. It cannot be stressed too often that TA critically depends upon obtaining knowledge – fact, conjecture and attitude – from wherever this knowledge may reside.

A form of workshop is the working party of a group of experts, who get together with the express purpose of reaching consensus and planning cooperation. The groups may be mixed, with experts from academic, government and commercial sectors, depending on their purpose. There is a lot to be gained by cooperation in the generation of a broad view of the future, leaving sufficient scope for individual firms to develop their own practical solutions and strategies. This form of cooperation among rivals, realizing that even rivals sit in the same boat, has become widespread and is particularly useful in gaining insights into what the future – or futures – might be, particularly as the members of such groups are instrumental in shaping the future. Governmental bodies often encourage this form of cooperation and develop programmes that run under names such as Technology Foresight and similar. A recent example is the British project Research Foresight and the Exploitation of the Science Base, which attempts to stimulate collaboration in foreseeing useful avenues of future technological developments (Cabinet Office 1993).

Large corporations use similar techniques, essentially based on the exchange of opinions among experts and managers of departments and functions concerned. The exchange of opinions can occur via a messenger: somebody who questions plenty of people on the same topic and transmits and coordinates these views. Sometimes such exchanges are formalized and the results incorporated into strategic planning. A recent publication describes a method used by BP under the title Roadmaps. What this boils down to is that everybody concerned with the formulation of the strategic plan is consulted and the results of the planning exercise are presented in easily understood graphic form (Barker and Smith 1995).

The holistic view of TA presents certain pitfalls. When experts are asked to give their views of future events, it needs to be borne in mind very carefully where their expertise lies. An electronics expert should be asked to forecast events in electronics; his or her views on political, social, or commercial developments must be treated as those of lay people, or, sometimes, as those of representatives of specific economic or social interests.

In all forecasting we must distinguish very clearly between the internal logic of technological developments leading to a range of futures, and the possibilities of external changes influencing the pace or direction of such developments. The path of science and technology depends not only on its internal logic. It depends strongly on social selection mechanisms, which can take many forms. There may be financial support for certain scientific topics or technological developments, to the virtual exclusion of other topics. There may be consensus among commercial firms on the future of some developments, causing them to cooperate, however loosely, in this area. There may be powerful currents of public opinion, possibly caused by pressure groups and enhanced by marketing strategies, in favour of certain developments. There may be changes in regulations that call for particular technological responses and developments. A technology develops in response to internal technological logic as well as in response to social forces. The market a technology finds also depends on the properties of the technology itself, but equally on social and economic circumstances, such as fashions, disposable income, and so forth. The team of analysts should integrate these various aspects, but must beware of confusing them.

The Delphi method can be employed more directly for technology assessment purposes. If we ask a suitable sample of experts – or even the population at large – what they think the most important problems related to the deployment of technology are and what solutions to these problems, technical or political, they think are possible, we are at the very core of technology assessment.

If such a method were to be employed by industry in the course of forming strategic plans for technology, the Delphi method could focus on issues relevant to this task. We could ask experts to reach consensus, if possible, on what product innovations are technically within the reach of the company. A different group of experts could be asked what chances they think the proposed innovations might have in the market. We might ask experts what production methods they see developing and which of these might be useful to the company. We might ask what environmental problems the experts regard as important and whether they foresee that the company might make contributions to their solution. In addition, we could ask what action the company needs to take to avoid falling foul of future environmental regulations. The above sketch shows possibilities of using the Delphi method creatively in the service of an industrial enterprise. It represents a systematic search for ideas and opinions from suitable experts and, at the same time, attempts to give weight and credibility to such ideas by reaching consensus on them.

We can attempt to reach consensus in direct group discussions. These have been termed *consensus conferences* in the realm of trying to obtain public views about technological problems. In the industrial context, this means preparing clear presentations of the issues at hand. These can either be distributed beforehand, or at an internal conference. At the first meeting questions can be raised for clarification, so that all participants will be well informed and under no misapprehensions. The next meeting should serve the purpose of reaching consensus. Speakers representing different interests and different opinions present their points of view. If the meeting is carefully chaired and the chair attempts to show exactly where the differences of opinion or interpretation lie, it may be possible to reach full or near consensus. It is the nature of democracy that consensus is not absolutely necessary, as the minority is willing to accept the majority opinion. In any event, this type of conference is useful for top management to achieve an awareness of the different streams of opinion, and of the majority opinion, and thus is helpful in forming top management's own view.

Modelling

Whereas extrapolation consists of extending real historical data for some variable into the future by fitting a curve to the data, modelling consists of substituting a plausible mathematical formula for real relationships between several variables. The mathematical equation serves as a model that simulates reality. Describing the behaviour of a single variable over time is the simplest type of model. Models can be much more elaborate. They can deal with a large number of variables and functional relationships between them. Time can be one of the independent variables. Models can use real data as a basis for postulating mathematical relationships, but they can also use plausible equations as a substitute for real data. The essence of a model is a collection of postulated mathematical relationships between a number of variables that substitutes for and depicts real functional relationships. The model is only as good as the approximation of the postulated functional relationships to reality.

In principle, it is perfectly feasible to construct models without the use of computers. However, because of the computer's capacity to handle a large number of equations with a large number of variables, it is eminently suitable for the construction of complex models and for the graphical display of the results of mathematical manipulations.

The great advantage of models is that they can introduce feedback loops. Thus a change in any one variable can affect changes in any other variables. We can therefore follow the changes in a variable caused by its dependence not only on one other variable, but on any number of cross-impacts. This makes the model more realistic and distinguishes it substantially from mere extrapolation.

Models are widely used for forecasting, as it is a simple matter to extend the variable time into the future, while maintaining all functional relationships and introducing feedback loops that represent additional functional relations. Two conditions must be fulfilled to obtain reliable results: the model must be reasonably accurate in depicting the functional relationships between the variables correctly, and the relationships must hold over the time period of the forecast. The best known examples of these kinds of forecasting models are meteorological models and economic models. As with all models, real data are used as inputs into the functional relationships. The real data can be historical and/or instantaneous. Thus in an economic model recent figures for variables such as prices, manufacturing output, balance of trade, money supply and interest rates are fed into the model. For weather forecasts, data from satellites, weather balloons and observatories are used to supply values of variables such as wind speeds, temperature and humidity. Both economic and weather forecasts are quite reliable over very short periods, but the models break down if the time horizon is extended.

A model that has achieved great prominence (or notoriety, depending on your point of view) among people interested in technology and the environment is the world model by Meadows et al. first published in popular form in 1972. The essence of the model is to show that if the use of natural resources continues to grow exponentially with time, then the limited reserves will, sooner or later, be exhausted. Similarly, if the ecosphere continues to be loaded with waste products at an exponentially increasing rate, it will eventually be unable to absorb them. Using the computer's capability of simultaneously dealing with a large number of variables, the model elaborates on the above theme and shows how, with different scenarios of use and conservation, the natural and economic systems will face some form of collapse. The model is extremely complex, with many non-linear relationship and elaborate feedback loops. It has been heavily criticized by critics equipped with hindsight for inaccurately depicting the then future. That, however, is missing the point. The model did show graphically that exponential growth is unsustainable in the long run and that various conservation measures can make a difference. And that was, and is, what mattered. As all good forecasting, the model provided a range of scenarios and provided options for actions that can shape the future. Some of the authors of the model have recently written a sequel to the original publication, in which they concentrate less on a description of the model and more on the burning issues of conserving resources and safeguarding the natural environment (Meadows et al. 1992).

Even complex engineering systems can be satisfactorily modelled and indeed computer models now replace much experimental work, such as wind-tunnel testing for the design of car bodies. The models are accurate because the equations connecting the variables are sufficiently well-known. These are not, of course, forecasting models, except in the sense that the model predicts how a certain engineering system will behave under certain circumstances. The system may be a motor car, or a chemical reactor, or a nuclear reactor, or any other physical system – we can predict its behaviour provided we understand the system sufficiently well. Though the weather system is a physical system, it is too complex and subject to too many forces to be accurately predictable. The situation for social systems is different. It is impossible to construct accurate models of complex social systems for the simple reason that mathematical relationships between the many variables are not known and probably do not exist. In systems where the freely exercised will of many human actors affects the results, no mathematical relationship can exist and even statistical relationships are of limited validity. Experience has shown that it is futile to try and construct models of great complexity and sophistication – relatively simple models, intelligently constructed, are as good at highlighting the essential features of a problem.

Although models of social systems, or systems in which humans interact with physical systems, can never be accurate, they nevertheless can be used to provide useful guidance for actions that affect the future.

Technology forecasting

So far we have made no distinction between forecasting in different disciplines, and have mentioned economic, environmental, meteorological and every other type of forecasting in a true kaleidoscope of futurism. This may have caused some confusion, which we shall try to remedy now by concentrating on technological forecasting alone, although for a comprehensive technology assessment, technology forecasting alone is not sufficient. We may need at least market forecasts, and possibly economic, social and political forecasts, as additional inputs.

The development of technology has two characteristic features relevant to forecasting: technology normally develops by incremental steps along a trajectory that leads to an ever higher figure of performance; and major technological developments occur in known areas of greatest activity at any given time.

The figure of performance is a composite of the various main characteristics of performance of a given technology. A computer, for example, may be characterized by its price, by the speed of processing, speed of access to memory and the amount of memory available. An aeroplane may be characterized by its pay-load, its range, its speed, its fuel consumption, and its operating costs. An internal combustion engine may be described by its power-to-weight ratio, its energy efficiency, its exhaust characteristics and its longevity. We may use a single composite figure of performance or we may use individual characteristics of a given technology; in all events we can use historical data to foresee future developments by extrapolation. Two things must be borne in mind: first, most performance characteristics tend toward some limit given by laws of nature and the more highly developed a technology is, the more difficult does it become to achieve further progress. If we stay with the example of the internal combustion engine, the development of a new, improved, engine has become enormously costly and progress has become very slow. We call this the law of diminishing returns on R&D. Second, which characteristics of a technology are chosen for improvement and development is a matter of social selection environment. The catalytic converter, which greatly reduces environmentally harmful emissions from internal combustion engines, has been developed under pressure from legislation that arose out of new environmental concerns. Pure extrapolation would never have forecast this result; it could be foreseen only by looking at social forces that select certain technological features in preference to others. It illustrates yet again that extrapolation must not be carried out without regard to external circumstances that might change the direction of development. A radical technological development can similarly render extrapolations null and void. If the internal combustion engine were to be replaced by an electric motor, by a turbine or by a steam engine, all extrapolation of ICE features would become pointless.

Extrapolation becomes entirely useless when dealing with a radically new technology. In the absence of historical data to extrapolate from, the only guide to the future is the intelligent and informed guess, i.e. expert opinion. Expert opinion can, however, be wrong. For a long time many experts believed that the so-called Stirling engine – a form of heat engine – showed great promise for development into a highly efficient engine. A great deal of money and effort was invested into the development, but after some years it became obvious that perseverance with this work was turning into obstinacy. The promise remains unfulfilled, the obstacles proved insurmountable. Another case in point is the fusion reactor. Expert opinion has consistently predicted over the last twenty-five years that this will become a practical proposition in thirty to fifty years.

Dominant technologies

Technology forecasting is aided by knowledge of the most active technological fields, the dominant technologies. In some ways, technology forecasting itself helps to pinpoint the most active areas that will be the most dynamic and most important over the next few years. No matter which government or private report about important innovation over the next number of years one reads, they all emphasize the same technologies as dominating the future: information technology – with slogans such as multi-media and interactive; genetic engineering, stressing agricultural, medical and pharmaceutical applications; new materials, with emphasis on composites (such as fibre reinforced plastics) as materials for construction or car components and the possibility of entirely new materials for electronics, sensors, and computing; transport – because of the urgent need to solve the transport crises of all large cities; energy as the foundation of everything else and, as an also ran, environmental technologies which include renewable energy, low-consumption cars, high efficiency heating systems, better sewage and rubbish disposal.

This type of forecasting is meant to guess where the action in the next few years will be. In a way, these forecasts are self-fulfilling prophecies. If there is consensus on where the action shall be, this is where it will be. It is where the competitive battles will be fought and where – apart from military technologies – the main thrust of research and development will be. It is somewhat like fashion: you can't go wrong if you are fashionably dressed. Running with the pack gives a sort of safety in numbers, and the knowledge that these areas are dynamic opens up opportunities for innovative products as well as innovative applications. On the other hand, there may be less fiercely contested territories elsewhere, where it may be easier to make a living either by using old methods efficiently or by being innovative.

This type of forecasting plays the role of agenda setting. It helps to determine the dominant technological paradigm of the period. The present period has been characterized by information technology for some time, with various words such as multi-media and interactive being added more recently. Although new materials and genetic engineering are important – and are seen to be important – no other technology has either developed so fast or had such dramatic impacts on as wide a sphere of activity. Never before have so many lives been affected so dramatically by so few basic technological innovations. The numerous progeny that sprang from the union of computers with semiconductors have truly changed the world.

If there is consensus that certain technologies are in the forefront of development and action, then indeed action will concentrate on these areas. The trend tends to be reinforced by governmental activities: if an area is singled out as progressing fast and as being important, then governmental agencies tend to spend money and effort on supporting it. Research grants, investment aid, training schemes and other supportive activities are lavished upon what is regarded as key areas of growth. As the world industrial system has become closely interlinked, governments all over the industrialized world tend to support the same technologies, thus enhancing their speed of progress as well as the fierceness of competition.

Cost–benefit analysis

One of the commonly used methods of technology assessment is cost–benefit analysis. The principle is simple and unassailable. Given several options for a decision on technology, compute both the costs and the benefits of each option and, other things being equal, chose the option that has the best ratio of benefit to cost.

The devil, as so often, is in the detail. The two principal difficulties are (i) to assign financial values to costs and benefits that may be intangible, (ii) to measure costs and benefits without attaching non-financial preferences to them, and (iii) to value future costs and benefits correctly.

A moral dilemma revealed in cost–benefit analysis, as in much technology assessment, is that the costs are not necessarily borne by those who reap the benefits. In the example that follows, cost–benefit considerations for the location of an airport, the benefits accrue to the airlines and their passengers, whereas the costs are, to a considerable extent, borne by the local population affected by noise and disruption. In a technology assessment carried out for the sole use of a commercial firm, the costs and benefits are primarily those of the firm. But some costs may be environmental and some may have to be borne by the firm's customers. Though it is desirable for a firm to externalize as much of its costs as it can, it is becoming increasingly difficult to do so both for legislative and for image reasons. Certainly a technology assessment would be failing in an essential aspect if it did not show who would bear costs and who would reap benefits. More generally, when discussing impacts of a technology, the assessment must show where, and on whom, the impacts fall.

An historical example may illustrate some of these points. When it seemed obvious that the two existing London airports, Heathrow and Gatwick, could not cope with the expected (from extrapolation) increase in air traffic, a commission (the Roskill Commission) was set up by the British government in 1968 to report on the most favourable site for the construction of a third London airport. The commission used cost–benefit analysis in the most meticulous manner. Four short-listed sites – Cublington, Foulness, Nuthampstead and Thurleigh – were evaluated in pain-staking detail. The recommended site turned out to be Cublington. The government of the day eventually decided against building a third London airport. Instead, the two existing airports were expanded and, many years later, the existing small airport at Stansted was completely reconstructed and designated as the third London airport (with Luton as something like a fourth airport serving London).

The cost–benefit analysis encountered some truly bizarre prob-lems. Problem number one was the allocation of a monetary cost to the inconvenience caused by aircraft noise to householders in the vicinity of the airport. The study attempted to solve this problem by calculating the depreciation in property values caused by the proximity of the airport, basing its findings on experience at Heathrow and Gatwick. It also used a survey method, in effect asking householders a hypothetical question: what monetary loss would they accept in exchange for moving into a quieter area? The total cost of noise was thus the depreciation of property for those staying put, plus the hypothetical realized loss on the sale of their properties and removal expenses for those moving out of the area. Both costs are subject to great uncertainty. It is also doubtful whether people who live in houses affected by noise suffer only a depreciation in the value of their property – surely they suffer a substantial intangible loss in their quality of life.

Similar difficulties were encountered in valuing the time spent by air-travellers on their journey to and from the airport. What is the cost of one hour spent by one person on a train, in a car or on a bus, quite apart from the cost of the journey? Is it the loss of potential income by the self-employed; the cost of loss of leisure by the employed; the cost of loss of production to the employer or to the national economy?

Some fiercely controversial questions arose out of the impossibility of attaching a value to the loss of habitat for wildlife, or the attempt to attach a value to the loss of a Norman church by calculating the cost of rebuilding it on a different site. Is an ancient church rebuilt on a different site still the same ancient church? And, if not, what is the monetary value of the difference between them?

The last point illustrates both the difficulty of attaching monetary value to intangibles and, above all, the difficulty of accepting the accountant's verdict. For those who attach a high spiritual value to an ancient church in its ancient setting, the accountant's argument is unconvincing. It would be equally unconvincing to ask the parishioners what value to attach to their church, for such hypothetical questions cannot be given a reliable answer. And for those who are passionate about wildlife, the cost of destroying habitats verges on the infinite.

There is a well-known problem of attaching a value to a future cost or benefit. For if I am given a sum C_0 today, I can invest it and obtain interest i per cent on it, increasing the value over the years according to the equation $C_t = C_0 [1 + i/100]^t$ where C_t is the value of my capital after t years. If, on the other hand, I am given the same sum C_0 in t years, I have lost $C_t - C_0$. Thus a benefit now is worth more than the same benefit in the future and, similarly, a cost now is greater than the same cost in the future. The net present value (NPV) of a future cost C_t is thus $C_0 = C_t/[1 + i/100]^t$ and the net present value of a future benefit is calculated in the same way. This is known as discounting and the rate of discount is a hypothetical level of interest that is crucial to the outcome of the calculation. The unknown level of future interest rates makes discounting for the future somewhat arbitrary. An additional uncertainty is, of course, the unknown rate of future inflation.

Cost–benefit analysis is not only useful, it is indeed indispensable in the assessment of any project involving future costs and benefits, i.e. in virtually all investment decisions. It helps to decide between projects in that it shows which project can be expected to yield the greatest net benefit over the expected lifetime of the project. However, cost–benefit analysis should not be extended to the evaluation of matters that have no meaningful monetary value. The value of the natural environment, the value of ancient monuments, or the value of a human life cannot really be measured in terms of money. One might do such calculations to compare the outcome of different projects, but when all is said and done, intangibles remain intangible and decisions on them must be based more on political than on financial considerations. It is a sensible rule to measure only the measurable and to evaluate immeasurables by other value judgements.

Cross-impact analysis

In analysing the impacts of a technology on a complex system, we must remember that a primary change in any part of the system may have an effect on other parts of the system. Thus we need to know not only the direct impacts of the technology on various parts of the system, but also the impacts of the impacts. Indeed, the occurrence of certain events may alter the probability of the occurrence of some other events and a change in certain trends may affect other trends. This can, in principle, be shown by constructing a matrix which allows us to register cross-impacts. Wherever possible we can quantify the impacts, but very often we shall have to be content with statements such as weak, strong, negligible or the equivalent in a number system, say from five for strong to zero for negligible. It is also possible to allocate probabilities to the occurrence of events, though the reliability of the numerical probability values cannot be very high. If we use numerical probability values for all the matrix, then it can be further manipulated mathematically. However, it seems to me that these manipulations give a false sense of precision and that it is preferable to use a cross-impact matrix – if at all – only as an aid to gaining an overall view of what is influenced by what and how strong these influences are.

The principle of a cross-impact matrix is shown in Figure 5.2. Each element of the matrix shows the influence of the event or trend in the first column on the events or trends in the respective row.

Let us use as an example the impacts that may be expected from introducing a machine-readable card, to be carried by all citizens, recording their medical details. The example is not fully worked out and the trends and events listed only serve the purpose of illustration.

	E1	E2	En	T1	T2	Tj
E1						
E2						
En						
T1						
T2						
Tj						

Figure 5.2   A cross-impact matrix.

*Source:*After Hetman, 1973, 242.
Note: En is the nth event, Tj the jth trend.

Some possible trends that may be relevant:

1   People are increasingly mobile and thus change addresses and their general practitioners frequently.
2   Hospital emergency services are overburdened and understaffed.
3   A variety of severe allergies are becoming known and prevalent.
4   People with chronic ailments live longer.
5   Potent drugs that interact with other drugs are increasingly coming into use.
6   People are medically insured by a greater variety of insurers, not only by the National Health Service in Britain or its equivalent in other countries.
7   Computers are in general use in doctors' surgeries as well as in hospitals.
8   A great variety of machine-readable cards are on the market, some with considerable memories and good security arrangements.
9   The use of medical cards is becoming widespread.

Some of the likely relevant events are:

1   Medical cards become standardized and all medical practitioners and hospitals acquire the necessary hardware and software to read them and write onto them, while maintaining

the same confidentiality as with paper-based medical records. Even ambulance crews (paramedics) obtain access to such medical cards.

2   Medical practitioners are given the option of putting only vital information onto the cards, such as chronic illness, blood group, severe allergies, present medication, while keeping all other records on their own computer to be made available to other medical practitioners or hospitals only with the patient's consent or under legally prescribed special circumstances.

3   Computers become able to take direct dictation, so that medical records can be dictated and yet held on computer.

4   Computers become able to read complex hand-written notes.

5   The carrying of medical records – of the variety meant to ease emergency treatment only – becomes compulsory.

6   The carrying of identity cards becomes compulsory (as it is already in some countries).

The example (Figure 5.3) shows that the method of constructing a cross-impact matrix forces the analyst to think about some unexpected cross-impacts, but it is cumbersome and tedious and really useful only in a limited number of cases. In a recent technology assessment of the medical card, the method was not found useful and was not used (Peissl and Wild 1996, 334–354). The matrix is somewhat confusing, but it does show the areas of strongest influence and thus highlights them for further study.

The important questions in a TA of the medical card are not really shown in the matrix. They concern costs and benefits, relations between doctor and patient, and, above all, questions of possible abuses of the information on the cards. Cards that carry the full medical history of the patient are currently unthinkable – both on technical grounds and for fear of loss of the card and objections by patients to having truly sensitive data so readily accessible. Potential employers might, for example, demand to see such cards and information about mental health episodes, or HIV tests, or similar, might be revealed against the patient's wishes and interests (Peissl and Wild 1996, 343).

Relevance tree

A useful method of mapping the possible impacts of the introduction of a technology on various spheres is the relevance tree. The main 'branches' are the main spheres of impact, the smaller

	E1	E2	E3	E4	E5	E6	T1	T2	T3	T4	T5	T6	T7	T8	T9
E1		0	0	0	4	0	0	-2	0	1	0	0	4	-4	5
E2	5		0	0	3	0	0	-2	0	1	0	0	3	3	4
E3	3	0		-1	1	0	0	-2	0	1	0	0	3	3	3
E4	4	2	-1		2	0	0	-2	0	1	0	0	3	3	3
E5	5	3	2	2		0	0	-3	0	1	0	1	3	-3	5
E6	4	1	0	0	4		0	-1	0	0	0	0	0	0	5
T1	4	2	0	0	3	4		1	0	0	0	0	0	0	2
T2	5	3	2	2	5	1	0		0	-1	0	2	3	3	4
T3	4	3	2	2	3	0	0	3		-1	1	0	2	3	3
T4	4	3	2	2	3	0	0	3	1		3	1	1	2	3
T5	4	2	2	2	4	0	0	3	0	2		1	2	2	3
T6	4	2	2	2	4	0	0	-2	0	0	0		2	2	0
T7	4	2	2	2	4	0	0	-2	0	0	0	0		1	4
T8	1	1	1	1	2	2	0	-1	0	1	0	1	1		5
T9	5	4	3	3	4	1	0	-2	0	1	0	1	4	1	

Figure 5.3   A cross-impact matrix for the assessment of machine readable personal medical record cards.

Note: Influences of the event or trend in column one on the events or trends in the respective row are graded from 0 to 5, either positive or negative.

'branches' are the sub-divisions of the major sphere. Though the relevance tree does not tell us anything that cannot be described in words, suitably arranged in paragraphs and sub-paragraphs, it is useful for those who prefer a graphical representation to a purely verbal one.

Take as a simple example the introduction of a computer numerically controlled machining centre to replace a number of older machine tools, such as lathes and milling machines. The expected impacts to be considered can be listed something like this:

1 Economic

   1.1 Increase in total capital cost
   1.2 Increased output
   1.3 Savings in labour cost
   1.4 Increased maintenance cost
   1.5 Increased dependence on reliability of single machine

2 Technical/commercial

   2.1 Increased output needs to be absorbed, possibly by increasing total factory output if market permits, or by re-design of product or process
   2.2 Need for new factory layout and new transport system for parts
   2.3 Possible repercussions on several parts of manufacturing process
   2.4 Possible repercussions on suppliers of materials and tools
   2.5 Improved quality of production (fewer rejects)

3 Labour

   3.1 Number of skilled machine operators reduced
   3.2 Possible need to increase shift work
   3.3 Need to train operators for new machine
   3.4 Need for more trained maintenance staff (internal or external)

4 Personnel/social

   4.1 Need to re-deploy or make redundant surplus machine operators
   4.2 Need to cope with possible opposition
   4.3 Re-grading of new machine operators
   4.4 Possible need to re-negotiate several gradings, including maintenance staff

4.5   Possible need to negotiate shift work

5   Environment/health

5.1   Need to question health and safety hazards of new machine
5.2   Need to question environmental hazards caused by new machine

    5.2.1   by lubricants

    5.2.2   by waste materials.

Some of the same considerations, put in the form of a relevance tree, are shown in Figure 5.4.

This example, even though given in outline only, shows clearly the difference between traditional methods of assessing technology and contemporary technology assessment. Traditionally, only the suitability of the technology for the task in hand and its pay-back period (the time needed for the investment to pay for itself in terms of savings and/or earnings) had to be right. Considerations of the social impact of the technology, or its effect upon safety, or the environment, or the system of transport for materials and parts, were often left out of consideration. The results of such inadequate assessments were unpleasant surprises and the need for improvisation and fire-fighting.

The purpose of technology assessment is the avoidance of unpleasant surprises and, more important, the optimum use of technology. The optimum can be achieved only if the technology is viewed as part of a whole system of production, including all the hardware, software, people and environment in the broadest sense. Even the best machining centre is not much use if it causes friction in some other part of the manufacturing system, be it by parts piling up somewhere, or by causing strife among the workforce, or by drawing the wrath of the factory inspectorate.

Figure 5.4   Relevance tree for the introduction of a new machining centre.

6

SOME APPLICATIONS OF
TECHNOLOGY ASSESSMENT

Environmental impact assessment

The United States National Environmental Protection Act of 1969 made it mandatory for all major industrial or civil engineering projects undertaken by the state, or that required planning permission or other state involvement in the form of licensing or aid, to submit a detailed statement of the impact the project was expected to have on its environment. This is the environmental impact statement (EIS), which since has also become mandatory in the European Union and is one of the major tools of environmental protection.

The art of producing environmental impact statements is known as environmental impact assessment (EIA). It is a close relative of technology assessment, except that it is somewhat narrower in concept as it concentrates on a specific civil engineering or industrial project and on impacts in the immediate vicinity of the project.[1] Examples of projects requiring environmental impact statements are: building or enlarging a manufacturing plant, installing a sewage treatment plant, building a power station, building a major highway or an airport. The question of what constitutes a major project, and thus requires an EIS, has often been disputed, as has the question of what constitutes a deterioration in the 'quality of life'. In some cases of dispute on whether an EIS is necessary, the authorities can demand a preliminary assessment on which to base their decision on whether a full assessment is required.

The first question raised in an environmental impact assessment is the question of its scope. Scoping is controversial because the

---

1 For full details of environmental impact assessment (as well as technology assessment), albeit of older vintage, see the extensive textbook by Porter et al. 1980.

wider the scope, the greater the effort involved in producing the EIS and the greater the chance of finding something negative. Scoping is equivalent to the first step in a technology assessment, but is much more formalized because EIA is a procedure that is prescribed in great detail by the law. Scoping is done with the participation of the authorities that might be involved in the project and is often carried out with public participation. Indeed, the public may be highly interested, because if the parties representing opposition to a project are excluded during the scoping process, their case is more or less lost from the start.

At the end of the scoping process a draft environmental impact statement is produced that lists all the areas of impact that are to be further analysed. Public hearings are sometimes included in the process of gathering evidence for the draft EIA. This provisional statement describes the purpose of the project, alternatives to it, all the licences required for the realization of the project, the names of the authors of the provisional EIS, and a list of all the persons or organizations that are to receive the EIS. The provisional EIS is published and submitted to all interested parties for comment. After the statutory period allowed for comment and representation, the final EIS is published. This is normally a document of about 150 pages. A decision on the project is published at the end of a statutory period after submission of the EIS, and this includes all mandatory measures for environmental protection and means of supervising compliance with these decisions.

In the case of a public project, it is the authority that carries out the project that is obliged to produce the EIS. In the case of a private commercial project, the EIA may be carried out either by the sponsor of the project or by an approved firm of consultants.

The process of environmental impact assessment

The environmental impact assessment distinguishes between two phases of the project: the construction phase and the operating phase. Environmental disruption during construction is – or may be – entirely different from environmental disruption during operation. Although construction is only a temporary phase, it needs to be shown that all measures to ameliorate environmental damage and disruption to the life of the community will be taken. It also needs to be shown that the least disruptive method of construction has been chosen and that all incidental damage to landscape or environment will, as far as is possible, be made good at the end of the construction period.

Although the construction phase of a project may take several years and may be highly disruptive, it can only be justified in terms of the project itself and hence we shall address most of our remarks to the environmental impact assessment of the operational phase of the project.

Much as in technology assessment, the first step of an environmental impact assessment, once the question of scope is settled, is to describe the project and, of greatest importance, to justify it in terms of benefit or need. If, for example, a power station is to be constructed, the proposal must justify the project by showing that indeed additional power is needed.

Some form of cost–benefit analysis often forms part of the justification, clarifying particularly who the beneficiaries will be and who will bear the costs. Say the proposal is to build a road bridge. In that case the beneficiaries will be the users of the bridge, but the losers may be the landscape or wildlife (i.e. the public at large), or householders and shop-keepers who will be affected by the new flow of traffic.

The choice of project must be justified in terms of alternatives, i.e. by showing that this particular project is the most favourable of all possible alternatives. This can be shown by cost–benefit analysis of the alternatives. The so-called zero option (no project) is excluded by demonstrating the need for such a project.

The next step of the EIA is its very core, the actual assessment of environmental impacts. Not only the proposed project is assessed, but the alternative projects also have to be evaluated in terms of their environmental impacts. Mere economic superiority of any one project is not enough to justify a decision in its favour. One of the ways of looking upon the environmental impacts of a project is by regarding it as a system that interacts with its environment (see Figure 6.1). As the figure illustrates, we need to consider all the flows between the system and its environment. Incidentally, the construction phase needs to be analysed in much the same way, though the flows there will be entirely different from those during the operational phase.

Consider as an example some kind of manufacturing plant. Without claiming completeness, we may describe the inward flows as consisting of the inputs of parts and materials, including chemicals and lubricants, tools and machinery, energy in its various forms, water and food, information by telecommunications or mail, and

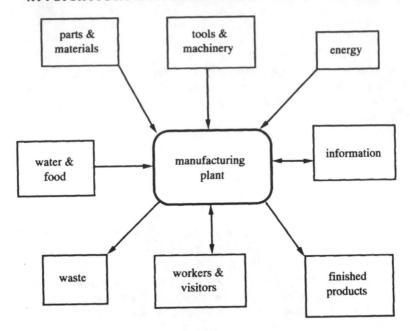

Figure 6.1   Inputs and outputs of a manufacturing plant.

the incoming workers and visitors. The outward flows consist of finished products, outgoing workers and visitors, outgoing information, solid and liquid waste, waste heat, effluents, and emissions.

The total environmental impact of the project consists of the once-and-for-all impact caused by its very existence, and of the dynamic impacts of the above flows. The effects caused by the mere existence of the plant may range from a substantial visual impact and a reduction in recreational/agricultural land, to an effect on wildlife, the road system and the local community, possibly by an increase in population and, it is hoped, by an increase in job opportunities. The impact caused by the operation of the plant can be analysed by looking at each of the flows.

The flow of materials and parts affects the environment by increased movement of lorries or trains. Some of the materials may be hazardous and special precautions may be needed for their transport. The movement of workers and visitors may cause increased local traffic, possibly on an already overloaded road system. It may be necessary to construct new roads – with all the difficulties and disruptions this entails – to ameliorate the traffic problem. The flow of information should not cause too many problems, except that the capacity of the local network may have to be increased. The same applies to the supply of energy, though this may entail the construction of new power-lines or the transport of large quantities of oil or coal. The supply of water may cause problems, as there may be a scarcity in the area and new sources of supply may have to be provided. It may be necessary to demand an efficient re-circulation system within the factory so as to reduce total consumption to a minimum. Because of the overall scarcity of clean water, it is a useful principle to demand this even if there is no present local scarcity.

The most problematic issues are wastes in their various forms. Waste heat may be a major problem, unless the site is by the sea and plenty of cooling water is available. Otherwise, it may be necessary to erect unsightly cooling towers or there may be a danger of disrupting the varied life of a local stream or lake by heating its water by the waste heat from the factory. On the other hand, it might be possible to use the waste heat for fish farming or for domestic heating. Emissions may be toxic and may have to be filtered, or, in the most harmless of cases, they may contribute to the greenhouse effect and may have to be reduced to the unavoidable minimum. Though noise and smell are not generally regarded as emissions, these may nevertheless cause inconvenience and may need measures to suppress the nuisance. Effluents may be very problematic indeed and strict cleaning and filtering procedures may have to be imposed or else the manufacturing process may have to be altered to reduce the total amount and/or the toxicity of the effluents. Even solid waste may be highly problematic. It too may be toxic and its disposal may be fraught with difficulty. Or else it may be bulky with nowhere to go and special measures for recycling, burying or burning it may be called for.

The EIS must not only discuss all these issues – or other issues that may arise in a project of a different kind – but must also show that the best possible solutions and methods of amelioration have been worked out and have been included in the project. It must also show whether the best possible is indeed good enough.

Environmental auditing

For the assessment of major products and production processes it is recommended to conduct an environmental audit. This means following a product from the cradle to the grave, counting at each stage of its life the inputs of materials and energy required and the outputs in terms of waste and environmental impacts of every kind produced.

Take, for example, a domestic washing machine. It requires certain amounts of metals, paint, rubber, plastics, electronic components, and energy for its production. These inputs can be followed back further, right to the initial sources of raw materials and to all the stages of processing and transporting them. The machine is then shipped to a customer and, during its service-life, consumes certain amounts of water, electricity, washing powder, other chemicals such as fabric softeners, and spare parts. All the water consumed is discharged into the sewers and has to be treated. This effluent and its impact on the environment can also be followed to its final destination. When the machine reaches the end of its useful life, it will either be dumped in its entirety or partly dismantled and recycled, depending upon its construction and prevailing environmental regulations and conditions. In any case, we can obtain a complete picture of all inputs, outputs and impacts of this product.

Many of the figures used will only be estimates and averages, but nevertheless a reasonably accurate picture can be obtained. This in itself may be interesting, but it only becomes helpful if we compare different makes or types of machine and different methods of production and final disposal. If we do the comparison in standard categories, we can draw conclusions on how best to design the particular product, how best to manufacture it, and how best to dispose of it. There is no need, indeed no practical possibility, to go into every last detail of an environmental audit. With a little experience a practitioner can soon find which items really count and can forget the small change as long as the main figures are correct.

Some manufacturers are taking environmental auditing very seriously. This is not only good for their image, but it also enables them to use the least wasteful, and thus most economic, processes and designs. As consumers are becoming more aware of the importance of environmental considerations and are also becoming aware of the cost of waste to themselves, so a favourable environmental audit for a product will increasingly become a competitive advantage. Similarly, as environmental regulators become more active, a favourable cradle-to-grave balance for products and processes will become a necessity.

Environmental audits allow us to choose best designs for products, best designs for manufacturing processes and, last but not least, best methods of final disposal. The advantages of using environmental criteria to achieve optimum designs, processes, and methods of disposal accrue to the manufacturers, to the users of the products, and to humankind in terms of improved environmental quality, less wasteful consumption, and fewer problems of waste disposal.

The topic of environmental auditing raises the question of total efficiency. Efficiency, in general, is defined as the ratio of outputs to inputs. In other words, how much input do we need to achieve a unit amount of output. A manufacturing process needs a great number of different inputs, and thus we can define a large number of different efficiencies. Environmental auditing is interested in the life-time environmental efficiency, which itself is a compound of material efficiency, energy efficiency and pollution efficiency. The latter can be defined as the amount of pollution caused by the production of the machine, by an operational unit of the machine (e.g. the washing of one unit of weight of clothes) and by its final disposal.

Total efficiency may have many more components. Barbirolli (1996) distinguishes twelve different components of total technological and economic efficiency, ranging from materials cycle efficiency, to final product operational efficiency, to product volume efficiency, etc. He suggests a method of taking all the varied efficiencies into account to obtain an overall efficiency of a production process and/or a technology. The method is too complex to be described here and, I suspect, to be useful in practice. A simplified calculation and definition of the main components of overall efficiency may, however, be useful and the concept should be borne in mind.

Resolution of conflict and participation

The environmental impact of many major projects is highly controversial. This is unsurprising and practically unavoidable. Conflict arises out of genuine conflicts of interest as well as out of genuinely different cherished values.

There are those who genuinely believe that more and faster technological progress is good and beneficial; and there are those who think that humankind needs to slow down, to think where the journey leads us, and to apply technology in more thoughtful and discerning ways. There are those who believe that fauna and flora and all things natural are subordinate to human needs and that these can expand for ever; while there are those who believe that it is the duty of humans to be guardians of nature and that such guardianship is the only guarantee for the long-term survival of humankind. There are those who believe that problems of traffic congestion can be solved by building more roads; and there are those who think that the only real solution is a reduction in traffic. There are those who believe that material production and consumption should expand for ever as the guarantors of progress; while there are those who think that material consumption should level off and that salvation lies in the search for harmony among all living creatures.

At this level of generality, the conflict of beliefs is beyond resolution. At a more mundane level of actual projects both sides may move away from their lofty stance and be prepared to discuss and resolve practical issues. Thus the first task of assessors of projects, or conciliators of conflicts, is to seek to reduce the issues from the esoteric to a practical, down-to-earth level. At this level it may be possible to find common ground, and to reach a compromise between practical utility and environmental damage.

Conflicts of interest may also pose almost insurmountable obstacles. If a proposed road threatens to cut off a bit of my cherished garden and to engulf my house in noise and fumes, there is nothing that can be done to mollify my implacable opposition to such a scheme. If the conflict arises between a community and a few individuals, then processes of law can be invoked to force the individuals to accept compensation for their losses. If the conflict is between a community and a powerful interest, whether state or private, then either a compromise backed by compensation may be reached or, if the worst comes to the worst, the authorities may override the community interest. If the conflict is between roughly equal groups, then compromise or conciliation may be possible. If not, the project will fall by the wayside. These outcomes are not necessarily desirable, but they correspond closely to reality.

The important issue in all cases is that all interested parties should be given an opportunity to present their case and to be properly heard by assessors, inspectors or conciliators who are seen to be objective in the sense of not being associated with any party to the conflict. This is easier said than done, but practice should come as close as possible to this ideal.

One of the difficulties to be overcome is to decide who the interested parties are. This, as was mentioned above, is part of the scoping, or similar, process. If the interested parties are people directly affected by a project, then the matter is easy. But if nature, fauna and flora, the climate, the ground-water, and similar general interests are at stake, who is to speak for them? Usually pressure groups form to fulfil this task, although in an ideal world the state should be expected to be the champion of the broadest interests of society.

The above matters are clearly for public enquiries and similar procedures, so why are they discussed in a text on technology assessment? The reason is simple. The technology assessor must be aware of these issues and, if we are dealing with an environmental impact assessment, the potential conflicts must be considered and their merits examined. This should be done in preparation for a public enquiry, in place of an enquiry, in summing up an enquiry, or in summing up private hearings that form part of the impact assessment. It is imperative that all environmental impact assessments, indeed all technology assessments, should find ways and means of incorporating the views and the interests of those affected by the subject of the assessment.

As a different example, let us consider the environmental impacts of a bridge. The justification for this kind of project must follow from considerations of the regional pattern of traffic or, possibly, from a wider consideration of overall transport planning. The alternatives to such a project, apart from not building anything (the zero option), are to build a bridge in a somewhat different location, to build a narrower/wider bridge, to build a bridge for both road and rail or only one of these, to build a tunnel, to build a bridge of a different design.

The impacts of a bridge are, of course, mainly on the total road/rail system and, hence, on traffic flows. These have secondary effects on communities, householders, trade, general planning, land values, and so forth. There may be major impacts on the landscape – bound to be controversial – and on the habitat of wildlife. There may be impacts on the flow of the river, on fishing and on navigation. These may all have secondary effects.

A bridge may represent a very large investment and the construction phase may take several years. Hence, the impact of the construction phase cannot be neglected. There may be considerable local disruption caused by the movement of large amounts of materials and by changes in road layout. There will be large numbers of temporary construction workers who may have to be accommodated and who will spend at least some of their income locally. The injection of capital will have multiplier effects, though they need not all accrue to the local community.

When the bridge is completed, it will provide little direct employment, except in maintenance and, perhaps, in collecting tolls. Its main benefits will be savings in time for travellers and, sometimes, easing of congestion in the vicinity. It is difficult to put a financial value on time saved (or lost) in traffic. One method is to assume that each adult traveller earns an average hourly wage and each hour spent in travel is worth that wage. Alternatively, it is possible to estimate the output lost to the economy during the time spent travelling by considering the average productivity of the economy. All methods of calculation are rather artificial and have to make assumptions about the composition of the group of travellers – how many children, how many pensioners, how many holiday-makers (and how do we value their time?). It is probably best to consider simply travel-time saved by the bridge, without attempting to put a monetary value on it. To complicate the issue further, it is known that almost all road capacity provided will, sooner or later, be used to the full and the old congestion will return, thus cancelling initial savings. The detailed savings depend so much on local circumstances, such as changes in routes travelled, that it is impossible to add much in terms of generalizations.

One more item does need to be considered, even though it may not be easy to reach definitive conclusions on it. It is the change in total emissions caused by traffic. There may be a change in total fuel consumption and, hence, a change in the emission of carbon dioxide, an important greenhouse gas. There may also be a change in the more directly noxious emissions, such as hydro-carbons, soot, carbon monoxide, and nitrogen oxides. If a reduction of these owing to the bridge can be demonstrated, this is a positive contribution to the quality of the environment.

Planning applications

At the beginning of a planning application lies a decision to build, enlarge, or alter some edifice serving – in our case – an economic purpose. This may be a manufacturing plant, a road bridge, a warehouse or even a coal mine.

It is frequently a task for technology assessors to prepare the ground for a decision on locating an industrial project in a particular location. There is, of course, a step preceding this task: a decision in principle to invest in a new or enlarged manufacturing plant. If the location of the plant is to be considered internationally, then the list of factors that could influence the location may read something like the following, listed in random order:

1   Political and economic considerations: these include stability of government, stability of currency, currency regulations, tax regime, investment subsidies, other state or regional aid, general infrastructure, labour laws, standards of education, standards of living, involvement in the political life of a country, environmental and other regulations, balance between new and existing plants.

2   Labour considerations: availability of suitable labour, labour morale, labour attitudes, availability of required skills, level of wages, level of training, living conditions, educational facilities, sport and cultural facilities.

3   Resource considerations: availability of raw materials, water, power, transport infrastructure, waste (including emissions, heat and effluents) disposal, information infrastructure.

4   Network considerations: proximity, quality and availability of suppliers of components, of R&D facilities, of rivals and collaborators, of suitable services.

5   Market considerations: proximity of markets, state of markets, trading conditions, freedom of trade, customs regulations, state of competition.

Once a decision on a desired location has been made, the material used for preparing the decision can be used for preparing a planning application. A planning application consists of showing the interaction between the proposed project and its physical, economic and social environment. Thus a planning application is closely similar to an environmental impact statement, though the planning application puts somewhat more emphasis on economic interactions and on minute details of the proposed construction.

We may consider a new factory as interacting with three local systems: the built environment; the natural environment; and the social and economic system. An environmental impact assessment deals largely, though not exclusively, with the physical systems, both built and natural. Hence we shall now consider mainly the interaction with the social and economic environment. In addition to considering the material flows discussed earlier, we now have to consider financial flows.

Take the same example as before, the building of a new manufacturing plant. The building phase may be an important source of temporary employment and a boost to local business. Indeed, the injected capital is multiplied in its effect on earnings; each pound invested leads to a greater amount of available income because a proportion of the money paid out in wages becomes incomes for other people, who again spend a proportion on wages and so forth. Altogether, an injection of a certain amount of investment into the local economy will cause a greater amount to become available – this is known in economics as the multiplier effect.

During the operational phase of the plant, the main financial flows are as follows:

1 Money paid out in wages to employees, which boosts the local economy by a multiplied amount because much of it may be spent locally. It may also cause considerable savings in unemployment benefit and other social payments. Indirectly, it may save on costs associated with crime, which is causally and statistically related to unemployment. The wages may not only help local shops and trades-people, but may also boost local house-building and local services of all kinds.

2 The second most important flow is through local supplies. It is very likely that the new plant will buy many of the materials, parts and services it needs from local suppliers. Each major economic player is surrounded by a large number of smaller players, linked together by supply chains, to the mutual benefit of both parties and the community at large. This may enable the suppliers to employ more people and the expenditure will, yet again, be augmented by multiplier effects.

3 The factory benefits the community by paying rates and taxes. The employees similarly pay rates and taxes. The firm may act as a sponsor for local educational, cultural or sporting activities, and its workers may use such facilities and thus raise the general level of demand for community facilities. The cultural and recreational infrastructure of the region may thus be enhanced.

The amount by which the community benefits depends, of course, on how much the factory spends locally, rather than in some remote part of the country or abroad. The benefits obtained from employment also depend not only on the number of jobs provided, but also on their quality. Quality of employment is measured by the level of wages, job security, the number of highly-skilled people employed and, last but not least, the training provided or sponsored by the firm. It also depends on whether the firm does only simple work based on designs and R&D carried out elsewhere, or whether it is involved in high technology and highly-skilled and creative work.

The local benefits also depend on the quality of the work demanded from suppliers. If the factory requires no more than simple supplies, the benefits are restricted, but if there is cooperation on high technology supplies, the benefits are likely to spread throughout the community and help to strengthen the economic and technological base of the region.

It may be possible to construct a model of the economic interactions of the new plant with the local economy and the rest of the world. We can then perform a kind of local input–output analysis, considering all the inputs and outputs of the firm and tracing where they go and where they come from.

Environmental impact statements and planning applications, taken separately or jointly, bear very close similarity to technology assessments. They look at the reasons for undertaking the project and evaluate alternatives to it. They assess the impact of the project on the economy, the social fabric, and the built and natural physical environments. If it is a major undertaking, the assessment must look quite a bit into the future, as the project may be planned in several phases. The remaining difference between a planning application, including an environmental impact statement, and a full technology assessment, is that the former does not consider in detail the wider repercussions of putting the firm's products on the market.

We shall now consider, in mere outline, two examples of technology assessments. One is rather broad in scope and considers the use of agricultural plant products as raw materials for industrial production. It is based on a major TA undertaken in the public domain in Germany (Wintzer et al. 1993). Although this is a major public TA, a similar study might be undertaken by a firm seeking new sources of raw materials in an attempt to put production on an ecologically sound footing, or by an agricultural enterprise seeking new markets. The other is an imaginary assessment, concerned with the production and marketing of a solar heating system, that might be undertaken by a small firm in Southern Europe.

Agricultural products as renewable raw materials

This technology assessment was financed by the German government. It was carried out by a core team, supported by external consultants and their teams, who produced partial assessments. The project was also supported by an advisory board. The latter is important, as it offers not only expert advice to the assessors, but also serves the function of representing economic and scientific interests in the broader community. The board was composed of university and industrial members and helped the assessors not only with advice, but also with contacts to numerous sources of information.

The first step of this TA, as of all technology assessments, consisted of determining and agreeing the topic and the scope of the assessment. The scope is always determined by considerations of feasibility within a reasonable time-frame and by availability of financial resources, though in this case it covers practically all aspects of the topic that come to mind. The topic consisted of considering the possibilities and consequences of increasing and broadening the use of agriculturally grown plants as sources for the production of energy, oils, starch, chemicals, pharmaceuticals, timber and textile fibres. The study was spatially limited to Germany and temporally limited to three periods: feasibility in the near future, feasibility within fifteen years, and foreseeable potential for the period beyond fifteen years. In this last case, the question was mainly what could be done now to start developing potentially useful plant-product combinations for that period. The main emphasis of the study was on the future up to fifteen years.

The topics to be covered in the study were all feasible plant-product combinations and their market potential; their environmental impacts (on air, soil, water, habitats); economic aspects; employment aspects; effects on agricultural and social structures; effects upon international trade and politics; need for, and effects of, policy intervention. The study was to describe the present position as well as future perspectives. It was to devote equal care to the analysis of advantages and disadvantages, and of chances and risks. To any connoisseur of TA this brief must sound perfect.

The complete assessment consists of a series of related assessments, each studying a different group of plants and industrial products. In each case the plant-product group and the technology associated with the various products is described. The descriptions include the state of the art, as well as the state of research and development and the future prospects. The impacts, advantages and disadvantages are discussed in detail. In particular, the full range of environmental impacts is analysed, from the effects on soil erosion and the need for fertilizers and their effects, to the final use and disposal of the products. Energy balances are calculated and effects upon employment, land use, agricultural and industrial structures are considered. Market opportunities are considered, based mainly on projected production costs and on acceptance criteria. The acceptance of well-known plants and products is high. Though farmers are a little more reluctant to consider novel plants and products, no major acceptance problems are anticipated.

Finally, policy issues are analysed. The policies range from various forms of subsidies, given for the sake of environmental protection, to regulations requiring the use of biodegradable materials where non-biodegradable materials cause major environmental problems.

Summary of results

Scenarios were developed for the year 2005, assuming a range of quantities of biologically-based products. The quantities of products were related to the use and availability of agricultural land. For each quantity of products, the effects upon the environment, upon employment and upon the economy were analysed.

The scenario with least use of biological raw materials is obtained if the reduction of carbon dioxide and other measures for environmental protection are not given high priority. If the environment is given high priority and attempts to reduce carbon dioxide emissions are taken seriously, the high usage scenarios are obtained. Taking into account anticipated changes in agricultural use of land – because of the need to reduce food surpluses and the need for less intensive use of fertilizers – no shortage of land for growing industrial raw materials is foreseen. Increased use of plant-based raw materials would create a modest amount of additional employment. The reduction in carbon dioxide emissions would be very substantial, perhaps more than 30 million tonnes per year.

Bio-energy

Roughly fifty combinations of plants and energy products were studied. Both liquid and solid fuels can be manufactured from plants. The technically most advanced are the production of fuel oil from rape seed and the production of ethanol. Rape seed could make a substantial contribution in the near future. Solid fuels will need further technical development that might take ten to fifteen years. The products with the most immediate promise are fuel pellets produced from straw and from waste wood cuttings. Certain grasses and fast-growing wood may become useful for solid fuel production in the future.

Though rape seed oil can be used for heating or as diesel fuel immediately, it cannot compete economically against fossil fuels unless it is given tax advantages, or unless the price of crude oil rises substantially. The same applies to all other plant-based fuels, though those produced from waste materials (straw and wood cuttings) come closest to competitiveness. The market potential of all bio-fuels depends upon the extent to which carbon dioxide emissions are reflected in the tax on fuels. Thus the market penetration of biologically-based fuels depends on whether or not society is willing to pay for environmental benefits in terms of either subsidizing such fuels, or giving them tax advantages, or taxing carbon dioxide emissions. Under favourable circumstances, it is estimated that renewable energy could supply between 5 per cent and 10 per cent of total primary energy consumption in Germany in the year 2005. In 2030 it could contribute up to nearly a third of total consumption. Bio-fuels take the lion's share (about 60 per cent) of the total renewable contribution.

The main environmental advantage of plant-based fuels is that carbon dioxide is recycled and thus the total emission is substantially reduced. Other environmental problems associated with energy production are also eased. The calculation of environmental impacts must, of course, take into account the use of fertilizers and the full cycle of growth, production and burning. The total energy balance, i.e. net energy gained per unit of farmland, is obtained by similar calculations.

The reduction in carbon dioxide emission varies from 80 kg/MWh (megawatt hour) when natural gas is replaced by methanol or hydrogen from biological sources, up to 390 kg/MWh if coal is replaced by plant-based solid fuels. The reduction of carbon dioxide emission for the replacement of fossil fuel oils by plant-based oil is about 100 to 200 kg/MWh.

If subsidies are used to help bio-fuels to compete, then a redistribution of incomes from the exchequer and from oil companies to the farmers and producers of fuel occurs.

Chemical products

There is a large variety of plant-based products. The main groups are:

Oils Plant-based oils are widely used in industrial products, mainly for the manufacture of detergents and soaps. A high percentage of these oils are imported from tropical countries. Some research is aimed at replacing oils imported from the tropical zone by oils grown in the temperate zone. The desirability of such substitution is controversial, to say the least. By genetic engineering and other chemical processes it is possible to produce various fine chemicals from oils, but it is too early to assess the market potential and practical utility of such products. The production of hydraulic fluids and lubricating oils from plants may become important. The main advantages are environmental, as these products are biodegradable. They would also reduce carbon dioxide emissions by reducing the need for burning used lubricants. The price of plant-based lubricating oils makes them uncompetitive, unless they receive subsidies or tax advantages, or their use is made mandatory in some applications where their environmental advantages are crucial, e.g. in lubricating chain saws in forestry.

Starch The main traditional use of starch is in the production of paper and cardboard. It is possible to use it for the manufacture of a variety of chemicals, including plastics. These materials can be biodegradable and their full potential benefit to the environment can only be utilized in combination with some form of recycling. The environmental advantages, however, have to be balanced against environmental disadvantages connected with growing suitable crops (mainly potatoes, maize and wheat) and manufacturing these novel materials.

Timber Apart from the direct use of timber in buildings and in furniture, it is widely used for the manufacture of cellulose, used in paper production. Paper production is notorious for its environmental impacts and many cellulose products are imported because their production is prohibited in Germany for environmental reasons. Environmentally more benign production processes are being developed. Whether these methods will lead to import substitution is an open question because the paper industry is largely multi-national and the choice of production sites in multi-national corporations depends on many factors other than environmental compatibility.

Three other groups of plant-product combinations are discussed: sugar, flax, medicinal plants and herbs. Sugar is used as an industrial raw material for the production of organic acids and in the pharmaceutical industry, e.g. for the manufacture of antibiotics, but also for the manufacture of glues, polymers, enzymes and other products. By the year 2005 something between 7 000 and 12 000 hectares of sugar-beet might be grown in Germany for industrial (non-food) use.[2] Flax is used for textiles, but conditions for growing flax in Germany are not economically favourable. There are about fifty or more different types of medicinal plants and herbs that can be grown in Germany. If the trend toward 'natural medicine' continues, it is likely that the demand for such preparations will require something like 10 000 to 15 000 hectares of medicinal plants by the year 2005. Some technological developments might improve the quality and the yields of the products.

Imaginary technology assessment for solar heaters

As a further example, consider a small firm in a Southern European country that wishes to expand its range of products by modestly priced solar heaters for domestic hot water. The example is worked out in outline only: no actual figures or actual analysis, only chapter headings.

Subject and scope

The subject is restricted to the consideration of solar heaters for domestic hot water, mainly for one-family houses. The possibility of extending the range to heaters for swimming pools is to be included. The boundary conditions are (a) that the heaters must be modestly priced so as to achieve a mass market; (b) that the total investment needed must not be too large; (c) that the product should reach the market not later than two years' time. The technology assessment must be carried out by three people in about four months.

Description of technology

The technology of solar heaters is simple in principle, but there is a great deal of detail which determines the quality and price of the product. Decisive are the materials used for the actual heater and for the housing and glazing, the surface treatment of the heater, material and location for the hot water tank, insulation, method of fixing heater on roof or free standing, prevention of rapid furring in hard water areas. Heaters for pools have no tank, but must be of larger

---

2   1 hectare equals 0.01 square kilometres.

capacity and must be able to withstand the corrosive effects of the chemicals used for water treatment. All these matters are important for the efficiency of the heater, the cost of producing it, the ease of installation, and its longevity.

Many surface treatments, materials and methods of construction are protected by patents. In all cases a certain amount of know-how and experience are required. The analysis must include information about firms that possess the know-how and/or certain patents. The question of what new methods are being developed in research laboratories must also be addressed.

Impacts of the technology

Environmental   The main effect is savings in fuel, as solar heat comes free. Thus there is a welcome reduction in carbon dioxide and other emissions from conventional heaters or power stations. The environmental effects of the materials and processes used in construction must be evaluated and compared. They should be a factor in the decision about materials and processes to be used. The materials and processes must also be evaluated from the point of view of possible health hazards to workers and users. The environmental impacts and cost of final disposal of heaters at the end of their useful life must be considered. Longevity plays an important role in determining total cost and total environmental effects per unit of time.

Economic   The main economic benefits should accrue to the manufacturers, their suppliers, and the installers of the equipment. The householders purchasing the heaters should also benefit substantially, otherwise there would be no market. The net benefit to the householder depends on the price of the solar heater, its life-time, the number of sunny days, and the cost of conventional fuel or electricity. The losers are the suppliers of gas and electricity, and the exchequer, because of loss of tax revenue levied on gas and electricity. This item is somewhat balanced by value added tax levied on the heaters. Because during the winter months there is insufficient sunshine to guarantee an uninterrupted supply of domestic hot water, the solar heaters are likely to be used in addition to conventional heaters. This poses the question of electric capacity that has to be made available for the winter, but is not used in the summer. Fortunately this is not a problem in a hot country, because electricity demand for air-conditioning is high in the summer.

An important economic effect might be the substitution of free energy for imported fuels, and thus an improvement in the balance of payments. This might be further improved by import substitution by the solar heaters themselves and possibly some export potential for them.

Employment and skills The net employment effect in the economy can be calculated on the basis of market estimates. More important for the firm, however, is employment within the firm. The analysis must produce an estimate of the number of people who might find additional employment and for the mix of skills that would need to be recruited. Considerable positive employment effects might be experienced by plumbers and builders who install the equipment. The firm might have to provide training for the installers and perhaps introduce some system of 'approved installers'.

Market estimates The total market depends on the number of householders who might benefit from the solar heaters, including new houses being built annually. The market further depends, critically, upon the economic benefits to the householders and on the ease, or otherwise, of the installation.

Policy options

The firm must decide upon the technology to be used and upon the range of sizes of heaters to be marketed. It must also decide whether to offer pool-heaters immediately, at a later date, or not in the foreseeable future. Further important policy decisions are: contractual arrangements with firms holding patents and know-how; plans for quantities of heaters to be produced; plans for arrangements with installers; marketing strategies. Options for all these decisions must be analysed and presented in the TA.

The firm must attempt to influence public policy toward solar heaters. Some of the policy options in the public domain are:

1   Pay an initial subsidy to the firm in order to obtain the public benefit of a reduction in greenhouse gases, a contribution toward conservation of resources, and the benefit of improvement in the balance of payments. The initial subsidy might be gradually reduced, and even withdrawn, once the volume of production is sufficient to make the operation self-financing.

2   Pay a subsidy to householders buying the devices until such time that the price falls sufficiently to make the heaters an economic proposition.
3   Help in the marketing operation by advertising the benefits of solar heat compared to fossil fuels.
4   Help with R&D and licensing arrangements for the firm.
5   Do nothing.

In summary, it may be said that reasonably priced solar heaters provide great benefit to the environment and to their users. With the right technology and the right policies and marketing strategies, they can be a major success in sunny countries. By comparing and contrasting the success of solar heaters in several Southern European countries, it may be said that public and private policies play a crucial role.

Some remarks on TA in the field of bio-technology

Modern bio-technology, and especially genetic engineering, occupy a special position in the field of technology assessment.[3] The general public fears these techniques and has the uneasy feeling that humans are meddling in matters that are beyond their comprehension and beyond their legitimate domain. Somehow Man transgresses into forbidden territory when he enters the domain of genes and heredity; this is the domain of God for believers and of Nature for disbelievers. This fruit of knowledge is forbidden.

But it is not only at the mystical level that people fear these technologies. Some of the fears are perfectly rational and are coupled with equally rational questioning of the need for such technologies. To make any progress in the analysis, we must distinguish between different applications and see which of these are feared, which are accepted, and how we can come to terms with the real issues.

The main applications that we distinguish are:

1   Use of hormones and other bio-chemical methods to affect the performance of farm animals.
2   Genetic manipulation of plants to give them desired properties and their use in agriculture and food production.

---

3   For literature on this topic see e.g. Levidow and Tait (1990); Tait (1990); US Congress (1990, 1992); CEC (1994); DoE/ACRE (1994); Wessels et al. (1995); Brauchbar et al. (1996); Torgersen (1996).

3   Genetic manipulation of organisms, up to large animals. Applications range from agriculture to the production of spare parts for human surgery.
4   Production of pharmaceuticals.
5   Treatment of waste.
6   Analysis of human genes with applications ranging from detection of criminals, early diagnosis of genetic defects, diagnosis of propensity for certain diseases, to gene therapy (either for affected cells or in reproductive cells).

The treatment of farm animals with hormones, though not really genetic engineering (except that some of the hormones are prepared with the aid of genetic engineering methods), has aroused the most heated debate in the case of BST (Bovine Somatropin) (Burkhardt 1992). This hormone is used to increase the yield of milk in cows. At first, it was feared that the hormone would be passed into the milk and might damage those who drink it. After a great deal of effort, these fears were allayed. However, opposition on social, economic and moral grounds remains unabated, if unsuccessful. Why, people ask, should yields be increased when we have surplus milk anyway? Why should smaller farmers be displaced from the market by larger farmers using every scientific aid in the book? Why should cows be first denied natural pasture and natural movement and then be made to produce abnormally large amounts of milk? Why do the authorities always support the chemical companies and the large farmers? Why is more research effort not devoted to dealing with waste and with other real problems, instead of with a non-existent problem of milk yields?

Some governments demand that any new product or process of modern bio-engineering should be socially compatible. But what does socially compatible mean? In principle it means that no net burden imposed by the novel product should be unilaterally imposed on any one group. It should also mean that no products or processes are introduced against the wishes of the people and that the people should have the right to participate in the shaping of technology. But technology assessment can, at best, only show what the burdens are and who will bear them. Technology assessment in the public domain may, under favourable circumstances, also be able to gauge opinions on new technologies. But it is up to the political process to avoid unfair burdens on social groupings and it is up to the political process that the will of the people in matters technological, as in all other matters, is carried out. The trouble is that the will of the people is a compromise between many wishes, and that governments tend to be highly selective in deciding which wishes are important. Perhaps genetic manipulation is a subject on which opinion research might be enlightening. Deliberate attempts at achieving consensus might also usefully be made. Social compatibility should, ideally, mean that technology is used in the long-term interest of humanity and nature. But who is to be the arbiter of such compatibility? Technology assessment and democratic politics is as near as we can get to the ideal.

In the case of genetically manipulated plants the fears concentrate on the possibility that the new genes might, if the manipulated plants were transferred to open fields, spread to other plants, or that the manipulated plants might spread and become difficult to control. If the plants are to be used for food, the fears are that some health hazard, as yet undiscovered, might lurk in them. As a result of these fears the use of genetically manipulated plants outside the laboratory is strictly controlled and permission to move the plants onto open fields is granted only when the hazards have been properly assessed. Whether or not these safeguards will prove sufficient in the long run remains to be seen; for the moment no mishaps have been reported, but the public remains sceptical.

The regulations are designed to allay fears of immediate hazards directly attributable to genetic manipulation. Normal agricultural hazards, such as over-use of fertilizers, difficult to control weeds and pests, over-use of pesticides and herbicides, damage to soil and ground-water, are not included in the new regulations. It is frequently argued that genetic manipulation of plants is not so very different from selective plant breeding and hence no new hazards are to be expected. The counter-argument is that entirely new genes are being introduced and that indeed new risks are involved.

When assessing risks, and allowing residual risks to be taken, inevitably the question arises of whether these risks are worth taking. In other words, what is the goal that is worth taking the risk for? And this is the weakness of risk analysis in genetic engineering: the public does not believe in the goal and rejects taking any risk for the sake of achieving what it does not want to achieve. However, these feelings are not very strong and not very vociferous; the authorities are anxious to show that the risks are very small indeed; and requirements for food labelling, especially for internationally traded foods, are so weak that the public generally does not even know that the food it buys contains genetically manipulated plant material.

If the genetic manipulation of plants is unpopular, the manipulation of animals meets with horror. Particularly the manipulation of large animals is regarded with moral revulsion, though the dangers involved in the introduction into nature of genetically manipulated small animals and insects appear to be greater because, once released, they are beyond recall. The possibility of breeding genetically modified pigs to provide hearts for transplantation into humans has been mooted and is receiving a mixed reception. The debate, and the research, continue.

The use of modern bio-technology for the production of pharmaceuticals arouses little controversy. People seem to worry about the efficacy and the price of pharmaceuticals rather than about the methods of production.

The use of bio-engineering for the treatment of wastes would be highly welcome, but these applications appear to be in their infancy. There might be excellent market opportunities for clever innovations in this field.

It is the manipulation of human genes which, predictably, causes the greatest concern. The mere analysis and mapping of human genes meets with some opposition, but also with some approval. It all depends on what it is to be used for. The mapping itself does not meet much vociferous opposition, even though it is a mammoth task swallowing vast resources.

When such information is used to detect genetic defects in a foetus and, possibly, though that may be in the distant future, cure the defect, then approval is certain. If the information were to be used to replace defective cells in young babies, an operation not feasible as yet, this also would meet with approval. Currently, the only use that can be made of the detection of genetic defects in a foetus is as an aid to a decision about the termination of the pregnancy.

Approval turns into disapproval and downright fear when it is suggested that genetic information might be used by potential employers and by insurance companies to weed out bad risks. If a propensity to some disease is genetically determined and becomes known to an employer or an insurer, this might be used against the individual. Thus the basic principle of insurance, when premiums are (or should be) charged on the basis of statistical, rather than individual, information and everybody is given a (nearly) equal chance, would be completely overturned. Those with the greatest need for medical insurance or life assurance would be denied it. And what of the individual concerned? Is it acceptable that we should know in advance what fate has in store for us? On the other hand, some knowledge might be useful in enabling us to take suitable precautions.

Worse than all this is that genetic information might be abused for selective breeding of humans (eugenics). And we have seen what happened in recent history when one nation considered itself a master race and attempted to reinforce its 'superiority' by selective breeding of 'superior' humans and selective killing of 'inferiors'.

For the above reasons all genetic manipulation of reproductive cells is regarded with grave suspicion, even if some attempts are well-meaning and aim at the elimination of defective genes that cause severe disabilities.

General approval is given to the genetic equivalent of finger-printing. Anything that improves the chances of detecting and correctly identifying criminals is regarded as highly desirable.

These few remarks on technology assessment in the field of bio-technology, very short of the full treatment the topic deserves, should further illustrate some characteristic features of TA.

First and foremost, the detailed approach to technology assessment on any given topic depends crucially on the topic itself. For methods of technology assessment the slogan 'horses for courses' is very apt. Even the very general methodology: scope, technology, impacts, policy (STIP), though universally valid, needs to be slightly adapted for each individual assessment.

Each particular technology has a dominant theme, rousing hopes and fears, that needs extensive treatment in a technology assessment. For TA in bio-engineering, the dominant topic is risk and risk analysis. In plant-based raw materials it is environmental effects and economic feasibility. In telecommunications it is social effects and policy issues.

Although some of our examples were taken from the public sphere because, unfortunately, industrial assessments are not available because of confidentiality, they should nevertheless demonstrate the principles, the issues and the approaches that apply to all technology assessments, whether public or private.

A final word of advice: approach all technology assessments with an open and inquisitive mind, seek both factual information and opinions wherever you can find them, and present your findings concisely, clearly and without fear.

Technology assessment is not the magic formula to save the world from human folly, but it is a step in the right direction. If tackled with energy, TA can make a far from negligible contribution toward a wiser use of technology and thus toward a better world.

BIBLIOGRAPHY

Barbirolli, G. (1996) 'New Indicators for Measuring the Manifold Aspects of Technical and Economic Efficiency of Production Processes and Technologies', *Technovation*, 16, 7, 341–356.

Barker, D. and Smith, D. J. H. (1995) 'Technology Foresight Using Roadmaps', *Long Range Planning*, 28, 2, 21–28.

Bessant, J. (1991) *Managing Advanced Manufacturing Technology*, Oxford: NCC Blackwell.

Brauchbar, M., Binet, O., Wessels, H.-P., Büchel, D. and Hieber, P. (1996) *Biotechnologie und Lebensmittel*, Technology Assessment 17, Bern: Schweizerischer Wissenschaftsrat.

Braun, E. (1981) 'Constellations for Manufacturing Innovation', *Omega*, 9, 6, 247–253.

—— (1990) 'Policy Issues in the Development of Telecommunications', *Technology Analysis & Strategic Management*, 2, 3, 265–273.

—— (1995) *Futile Progress*, London: Earthscan.

Braun, E. and Macdonald, S. (1982) *Revolution in Miniature*, 2nd edn, Cambridge: Cambridge University Press.

Braun, E. and Wield, D. (1994) 'Regulation as a Means for the Social Control of Technology', *Technology Analysis & Strategic Management*, 6, 3 (special issue on regulations), 259–272.

Brooks, H. (1992) 'Technological Assessment: Risks, Costs, Benefits', *Advanced Technology Assessment System, Issue 9, Biotechnology and Development*, New York: United Nations, 8–14.

Burkhardt, J. (1992) 'On the Ethics of Technical Change: the Case of bST', *Technology in Society*, 14, 221–243.

Cabinet Office (1993) *Research Foresight and the Exploitation of the Science Base*, London: HMSO.

Cabral-Cardoso, C. (1996) 'The Politics of Technology Management: Influence and Tactics in Project Selection', *Technology Analysis & Strategic Management*, 8, 1, 47–58.

Cas, J. and Pisjak, P. (1996) 'Integrierte Breitbandnetze – Eine Technologie sucht ihren Markt' in G. Tichy (ed.) *Technikfolgen-Abschätzung in Österreich*, Vienna: Austrian Academy of Sciences, 135–176.

Coates, J. F., Mahaffie, J. B. and Hines, A. (1994) 'Technological Forecasting: 1970–1993', *Technological Forecasting and Social Change*, 47, 23–33.

Commission of the European Communities (1989) *Perspectives for Advanced Communications in Europe*, Vol. 1, Executive Summary, Brussels: CEC, Directorate XIIIF.

—— (1994) *Biotechnology Risk Control*, Brussels: CEC, DG XI.

Department of the Environment (1994) *Annual Report No 1*, Advisory Committee on Releases to the Environment, London: HMSO.

Dosi, G. and Orsenigo, L. (1988) 'Coordination and Transformation: an Overview of Structures, Behaviours and Change in Evolutionary Environments', in G. Dosi, C. Freeman, R. Nelson, G. Silverberg and L. Soete (eds) *Technical Change and Economic Theory*, London: Pinter, 13–37.

Drejer, A. (1996) 'Frameworks for the Management of Technology: Towards a Contingent Approach', *Technology Analysis & Strategic Management* 8, 1, 9–20.

Dussauge, P., Hart, S. and Ramanantsoa, B. (1992) *Strategic Technology Management*, Chichester: John Wiley & Sons.

Encel, S., Marstrand, P. K. and Page, W. (eds) (1975) *The Art of Anticipation*, London: Martin Robertson.

Freeman, C. and Hagedoorn, J. (1992) *Globalization of Technology*, Commission of the European Communities, Global Perspectives 2010: Vol. 3, Brussels: CEC FAST Programme.

Freeman, C. and Perez, C. (1988) 'Structural Crises of Adjustment: Business Cycles and Investment Behaviour' in G. Dosi, C. Freeman, R. Nelson, G. Silverberg and L. Soete (eds) *Technical Change and Economic Theory*, London: Pinter, 38–66.

Freeman, C., Clark, J. and Soete, L. (1982) *Unemployment and Technical Innovation*, London: Pinter.

Galbraith, J. K. (1974) *The New Industrial State*, 2nd edn, Harmondsworth: Pelican Books.

Graves, A. (1991) 'Globalisation of the Automobile Industry' in C. Freeman, M. Sharp and W. Walker (eds) *Technology and the Future of Europe*, London: Pinter.

Green, K., Jones, O. and Coombs, R. (1996) 'Critical Perspectives on Technology Management: An Introduction', *Technology Analysis & Strategic Management*, 8, 1, 3–7.

Head, S. (1996) 'The New, Ruthless Economy', *Prometheus*, 14, 2, 195–206.

Hetman, F. (1973) *Society and the Assessment of Technology*, Paris: OECD.

Howells, J. and Wood, M. (1993) *The Globalisation of Production and Technology*, London: Belhaven Press.

BIBLIOGRAPHY

Jahoda, M. (1992) *World within Worlds*, Commission of the European Communities, Global Perspectives 2010: Vol. 4, Brussels: CEC FAST Programme.

Keller, W. W. (1992) 'Institutional Structures and Technology Policy' in G. C. Bryner (ed.) *Science, Technology, and Politics*, Boulder: Westview Press.

Levidow, L. and Tait, J. (1990) *The Greening of Biotechnology: From GMOs to Environment-Friendly Products*, Milton Keynes: The Open University, Technology Policy Group, Occasional Paper 21.

Macdonald, S. (1995) 'Culture and Image in International Strategy: Engineering Myth and Metal-Bashing', *Technology Analysis & Strategic Management*, 7, 4, 355–369.

Mansell, R. (1994) 'Strategic Issues in Telecommunications: Unbundling the Information Infrastructure', *Telecommunications Policy* 18, 588–600.

Meadows, D. H., Meadows, D. L. and Randers, J. (1992) *Beyond the Limits*, London: Earthscan.

Meadows, D. H., Meadows, D. L., Randers, J. and Behrens, W. W. (1972) *The Limits to Growth*, London: Pan Books.

Medford, D. (1973) *Environmental Harassment or Technology Assessment?*, Amsterdam: Elsevier.

Mintzberg, H. (1994) *The Rise and Fall of Strategic Planning*, Hemel Hempstead: Prentice Hall International.

NISTEP (1992) *The Fifth Technology Forecast Survey*, Tokyo: National Institute of Science and Technology Policy.

OECD (1992) *Technology and the Economy*, Paris: OECD.

POST (1991) *Relationships Between Defence and Civil Science and Technology*, London: Parliamentary Office of Science and Technology.

Peissl, W. and Torgersen, H. (1996) 'Internationale Institutionalisierungsformen von Technikfolgen-Abschätzung' in G. Tichy (ed.) *Technikfolgen-Abschätzung in Österreich*, Vienna: Austrian Academy of Sciences, 30–66.

Peissl, W. and Wild, C. (1996) 'Patienten-Karten: Eine Technikfolgen-Abschätzung einer Anwendung der Informationstechnologie im Gesundheitsbereich' in G. Tichy (ed.) *Technikfolgen-Abschätzung in Österreich*, Vienna: Austrian Academy of Sciences, 334–354.

Porter, A. L., Rossini, F., Carpenter, R. A. and Roper, G. (1980) *A Guidebook for Technology Assessment and Impact Analysis*, New York: North Holland.

Porter, M. E. (1985) *Competitive Advantage*, New York: The Free Press.

Potter, S. and Hinnells, M. (1994) 'Analysis of the Development of Eco-labelling and Energy Labelling in the European Union', *Technology Analysis & Strategic Management*, 6, 3, 317–328.

Rakos, C. (1996) 'NFF Nebenstrom-Feinstölfilter Technikbewertung einer Abfallvermeidungstechnologie' in G. Tichy (ed.) *Technikfolgen-*

162

BIBLIOGRAPHY

Abschätzung in Österreich, Vienna: Austrian Academy of Sciences, 266–274.

Rhodes, E. and Wield, D. (eds) (1996) *Implementing New Technologies*, Oxford: Blackwell.

Rothwell, R. and Gardiner, P. (1989) 'The Strategic Management of Re-innovation', *R&D Management*, 19, 2, 147–160.

Scarbrough, H. and Corbett, J. M. (1992) *Technology and Organization*, London: Routledge.

Tait, J. (1990) *Biotechnology – Interactions between Technology, Environment and Society*, Milton Keynes: The Open University, Technology Policy Group, Occasional Paper 20.

Thomas, P. (1996) 'The Devil is in the Detail: Revealing the Social and Political Process of Technology Management', *Technology Analysis & Strategic Management*, 8, 1, 71–84.

Torgersen, H. (1996) 'Zwischen Risikoabschätzung und Sozialer Verträglichkeit' in G. Tichy (ed.) *Technikfolgen-Abschätzung in Österreich*, Vienna: Austrian Academy of Sciences, 67–104.

Ungerer, H. and Costello, N. (1988) *Telecommunications in Europe*, Luxembourg: CEC.

United Nations (1992) *Advanced Technology Assessment System, issue 9, Biotechnology and Development*, New York: United Nations.

US Congress, Office of Technology Assessment (1990) *Critical Connections: Communication for the Future*, OTA-CIT-407, Washington, DC: US Government Printing Office.

—— (1990) *Genetic Monitoring and Screening in the Workplace*, OTA-BA-455, Washington, DC: US Government Printing Office.

—— (1992) *Cystic Fibrosis and DNA Tests: Implications of Carrier Screening*, OTA-BA-532, Washington, DC: US Government Printing Office.

Utterback, J. M. (1994) *Mastering the Dynamics of Innovation*, Boston: Harvard Business School Press.

Wessels, H.-P., Fischli, A., Mathys, P. and Brauchbar, M. (1995) *Einfluss der Biotechnologie in der Milchproduktion und Verarbeitung*, Technology Assessment 7a, Bern: Schweizerischer Wissenschaftsrat.

White, B. L. (1988) *The Technology Assessment Process*, New York: Quorum Books.

Whittington, R. (1993) *What is Strategy – and Does it Matter?*, London: Routledge.

Wilkinson, B. (1983) *The Shopfloor Politics of New Technology*, London: Heinemann Educational Books.

Wintzer, D., Fürniß, B., Klein-Vielhauer, S., Leible, L., Nieke, E., Rösch, C. and Tangen, H. (1993) *Technikfolgenabschätzung zum Thema Nachwachsende Rohstoffe*, Münster: Landwirtschaftsverlag.

Womack, J. P., Jones, D. T. and Roos, D. (1990) *The Machine That Changed The World*, New York: Rawson Associates.

INDEX